Soil Conservation

3rd Revised Edition

NORMAN HUDSON

1st Reprint in India, 2015
(Limited Copies)

Reprinted by:

United Book Prints
4806/24, Ansari Road,
Daryaganj,
New Delhi - 110 002
Tel.: 011-41511055

ISBN: 978-81-926860-3-5

Printed in India

Contents

Acknowledgements

My grateful thanks are due to the many colleagues who responded to my requests for ideas, information, and recent publications for use in the preparation of this revised edition. The capacity for instant communication by phone and fax has changed the process of information exchange with friends and colleagues out of all recognition since the first edition. They are too many for their contributions to be individually mentioned, but they were most welcome, and I apologise for not being able to use all the ideas offered.

I thank the Rockefeller Institute for allowing me the privilege and pleasure of working on this edition as a scholar in residence at the Bellagio Conference and Study Centre at the Villa Serbelloni, which provides a wonderful opportunity for concentrated study in a most gracious setting. I am greatly indebted to Linde Ovington-Lee for her invaluable assistance with word processing the text, and to Thelma Nye for sympathetic and helpful editing of all the editions and revised reprintings over 25 years.

Ampthill, Bedford 1995 *N. W. Hudson*

The author and publishers thank the following for permission to include their photographs, and data and diagrams on which the tables and figures in this book are based:

Aerofilms Limited for plates 1.1, 2.5 and 10.11; V. Austin for plate 10.15; BEMIS Company, Missouri, for plate 13.9; Conservation Commission of Northern Territory of Australia for plates 13.15 and 13.16; CRC Press for plate 10.5(b); R. M. Dixon for plate 13.5; Earth Island for figure 1.2; Edward Arnold for figure 1.1; H. Elwell for plate 10.9; R. Evans for plate 8.1; Fairchild Surveys for plate 2.1; FAO for table 1.5, figures 6.4, 8.1, 8.2;

H. Hurni for plates 10.7 and 10.28; ICRISAT for plates 10.13, 11.13, and 11.14; ILCA for plate 10.14; International Soil Reference and Information Centre (ISRIC) for tables 1.2 and 1.3; H. Lal for plates 10.4 and 10.33; R. Lal for plate 8.9; C. Matzon for plate 10.32; A. Mitchell for plate 8.8; National Institute of Agricultural Engineering for plate 2.2; National American Space Authority (NASA) for plate 5.1; OXFAM for plates 13.6, 13.7 and 13.8; Peace Corps for figures 10.9 and 10.11; Philippine Tourist and Travel Association for plate 10.3; I. Pla Sentis for plate 11.1; C. Reij for plate 10.5; R. J. Rickson for plates 8.20 and 8.21; River and Sea Gabions Ltd for plates 13.18 and 13.19; T. F. Shaxson for plates 10.1, 10.6, 12.5, 12.6 and 13.3; Soil Conservation Service of New South Wales for plate 2.6; S. Sombatpanit for plate 8.2; N. Swanson for plate 8.18; United States Department of Agriculture for plates 4.1, 10.19 and 10.24; H. Vogel for plate 10.17; Western Australia Department of Agriculture for plate 13.4; Worldwatch Institute for figure 1.3; World Neighbors for plates 10.25 and 14.2; World Resources Institute for tables 1.1 and 1.4.

Preface

In this third edition the basic structure has been retained, and the main changes are bringing in ideas and experiences which have arisen since the second edition. For the benefit of readers who are familiar with the previous edition, and wish to know where the changes are:

- We have left out the discussion of wind erosion (old chapter 14) because that is now well covered in other texts.
- The detailed account of Land Capability Classification has been shortened (old chapter 9) because the trend is away from blueprint solutions and towards national or local planning procedures.
- The detailed discussion of the Universal Soil Loss Equation (old chapter 10) is replaced by a wider discussion of modelling in the new chapter 7.
- The reduced emphasis on structures for the control of erosion is reflected in the new chapter 10 which replaces previous chapters 6 and 8. The material on hydraulic design is nowadays more appropriate in design manuals.
- The increased importance of control through land husbandry results in a completely rewritten chapter 11 on biological control measures, and the questions of policy and implementation (old chapter 16) are discussed more fully in the new chapter 14.
- All chapters have been revised with updated references.

An important development during the last ten years has been new thinking about soil conservation policies, and strategies for their implementation. That is not fully discussed here because it is the subject of *Land Husbandry* (Batsford 1992). Some of the topics in *Land Husbandry* have been introduced into this edition of *Soil Conservation*, and readers will forgive some duplication, which is necessary when trying to reach two different audiences. *Soil Conservation* was written as a teaching text, and has evidently proved acceptable for this purpose as it has been in print for 24 years, with ten printings, three translations, and three editions. *Land Husbandry* seeks to reach a wider audience including managers and decision makers in governments and development agencies, and practitioners in the many disciplines in the care and development of land resources.

Ampthill, Bedford 1995 *N. W. Hudson*

1

Man and soil erosion

1.1 The relation between man and the earth's resources

The balance between the demand which a community of plants or animals makes on its environment and the ability to satisfy those demands is not the static equilibrium of a laboratory balance with equal weights on either side. It is more like the unstable balancing of a circus acrobat – a series of swings and overcorrections, resulting in an oscillation on either side of the true balance point. Thus a species of wild life increases in number according to the availability of its food supplies, but then goes beyond the optimum number. The natural correcting factors of starvation or migration reduce the population, but there is an overcorrection, and soon the cycle starts again with increasing numbers.

1.1.1 The Malthusian Thesis

In primitive societies the human population also oscillates about a mean as the limiting factors of starvation, disease, and war, maintain an uneasy balance against the natural tendency to increase. A serious imbalance arises when man learns how to modify the limiting factors, but allows the natural increase to go unchecked. The essence of the Malthusian Thesis is that the resulting instability is the inevitable order for mankind. The thesis is named after the English political economist Thomas Malthus who first gave it formal expression in MALTHUS (1966). In an age when significant improvements in health, hygiene, and the standard of living were just becoming possible, Malthus aroused great argument when he pointed out that these very improvements could lead to misery through overpopulation. The key point of Malthus's argument is that population increases in what he called *geometric ratio*, what today we would call an exponential curve, because whenever there is an increase in population there are more people to produce the next generation, so the rate of population growth accelerates – in modelling terms a positive feed-back loop. On the other hand, when we increase food production there is no feed-back loop, the best we can hope for is for a constant linear increase. The natural sequence is therefore for food demand to increase faster than food supply, leading at some point to food shortage.

In the last century the opening up of extensive areas of the unexplored world disguised the growth rate, and at a time when it appeared that the supply of new lands was inexhaustible, Malthus's ideas did not seem very relevant. Today it is clear that the world population is indeed increasing rapidly, and will continue to do so even if the growth rate declines from the rate of the last 20 years. Also the increase is greatest in the developing countries. Europe in the last century saw major changes with the diversion of labour from agriculture to industry, and a corresponding movement from country to town, accompanied by a rapid increase in population. This pattern is now appearing in many developing countries.

Malthus also pointed out that in nature certain checks come into play to curtail excessive population growth, such as malnutrition, starvation, disease, and migration. But while these limiting forces still largely apply to animal populations, man has learned how to avoid or overcome them, particularly by improved health care which has led to a dramatic decrease in infant mortality and a significant increase in longevity.

Once demand for products from the natural resource base, such as food, clothing, shelter, and fuel, exceeds the sustainable production, then we pass from the stable situation of living off the sustainable interest from the natural resource base, to the unstable downward spiral where we use up capital resources to meet the demand. In a review of global degradation of natural resources Lester Brown sums up the position as: 'Every country is practising the environmental equivalent of deficit financing in one form or another' (WORLDWATCH INSTITUTE 1993).

There is some evidence that in some countries at least the rate of increase of population is showing signs of slowing down. But even if this trend were to be maintained, its effect would be completely overshadowed by the fact that populations will inevitably continue to rise because of the huge numbers of children in developing countries who will become breeding adults within the next ten years, as shown in figure 1.1. In an analysis of the relationship between food production and population growth, PEREIRA (1993) finds it regrettable that the UN Conference on Environment and Development (UNCED) at Rio de Janeiro in 1992 failed to 'recognise and address the grim synergy between rapid population growth, increasing poverty, and environmental degradation'

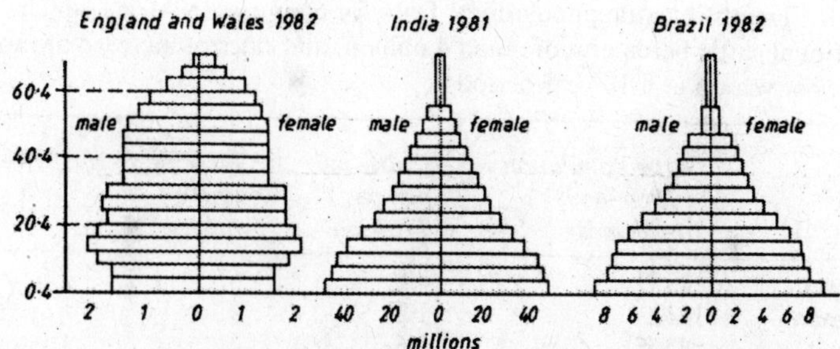

Figure 1.1 Age distribution of some populations. Each bar represents an age bracket of four years, with 0.4 at base of each pyramid (LOWRY 1986)

and also that 'the unsustainable rate at which expanding populations o: subsistence communities are destroying their environments was ignored'

There is a hypothesis, first presented in 1965, that population pressure is a significant factor in the intensification of agriculture and the adoption o: improved farming methods (BOSERUP 1993). There is some evidence tha in some cases population increase has been accompanied by increasec food production, the best example being the carefully researched anc documented study of events in the Machakos District in Kenya over the last 30 years during which the population has increased five-fold but the 1930's scenario of severe degradation and extreme poverty has beer replaced by greatly improved standards of living and of agricultura production (TIFFEN *et al* 1993). However, this does not mean there is necessarily a causal relationship between the change in population and the change in agricultural production which could have been because of the population increase or in spite of it. The Boserup hypothesis is no supported by current scholars OHADIKE (1992) and VAIDYANATHAN (1992)

On the other hand there is little doubt that in many developing countries the livestock population is rising at least as fast as the humar population, and in some countries at a faster rate (table 1.1) and that this is leading to severe degradation by overgrazing in many countries.

1.1.2 The exploitation of resources

For years there has been concern about whether the natura resources of the world are able to feed, clothe, and provide for the presen and future population. There is a wide range of opinion on this. The optimists believe that the potential for increase in food supplies is ample for foreseeable needs (REVELLE 1976), or that the problems are over-exaggerated (MADDOX 1972), or that improved farming methods car

Table 1.1 Trends in cattle populations. Data for countries in Africa which have national cattle herds of more than 4 million, and rates of increase of more than 2% per year over a 10-year period

	Cattle population in 1984–86 (thousands)	*Cattle per person 1984–86*	*Increase from 1974–76 to 1984–86 (%)*
Cameroon	4059	0.4	57
Madagascar	10423	1.0	22
Mali	4705	0.6	21
Sudan	22037	1.0	50
Tanzania	14031	0.6	24

World Resources Institute, 1988, chapter 17

increase yields (BOERMA 1975). The pessimists, nicknamed 'the Doomsters', who appear to be the stronger voice in current debate, believe that the demands on the resource base cannot be supplied indefinitely. In 1972 a computer simulation model was used to explore in *Limits to Growth* the effect of alternative trends in population, food production, pollution and so on, and came to the conclusion that whichever of the possible projections actually takes place, the world's resources will be inadequate (MEADOWS *et al* 1972), as shown in figure 1.2. Running an improved model in 1992 with 20 years additional data the result is that the conclusions are still valid but that: 'Human use of many essential resources and generation of many kinds of pollutants have already surpassed rates that are physically sustainable' (MEADOWS *et al* 1992).

Two key issues in this debate are how much the land base has been reduced by degradation, and how much it could be increased by the development of land presently unused. Degradation of the land resource is relatively recent in historical times. For countless centuries the world's capital of natural resources, minerals, the forests and the soil, was only required to yield a very modest interest which was sufficient to provide for man's requirements. As the population grew, it began first to extract an ever-increasing interest, and then to start using up the capital resources, likened to deficit financing by Brown (WORLDWATCH INSTITUTE 1988). In the latter half of the previous century and the first half of this, the industrial revolution and mechanization of agriculture put into man's hands the tools, the machines, and the engines, which made possible the exploitation of natural resources on a scale previously unimagined. This exploitation of non-renewable resources still continues. The large scale export of cereals from North America or beef or forest products from

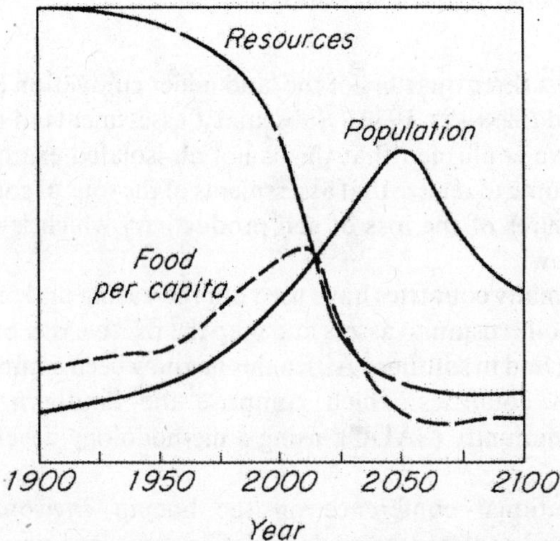

Figure 1.2 Computer simulation models attempt to predict the effect of possible changes. This is the prediction of what would happen if there are no changes in present trends of population, food production, pollution and so on (Adapted from MEADOWS *et al* 1972)

South America, or dairy products from New Zealand are all non-sustainable exploitation of land resource capital, achieved only by the using up of fertility, and accompanied by a decrease in the long term productivity of the land.

The food surpluses caused by the absurd Common Agricultural Policy (CAP) of the European Community are also non-sustainable. Some of the high yields are achieved at the expense of reduced long-term productivity, others by the high rates of inorganic fertilizer which is increasingly causing chemical pollution of water supplies.

In the second half of this century another threat has emerged – the increasing pressure on the land caused by the growing population, leading to the increasing use of land which is at best marginal and at worst quite unsuitable for sustained production. The problems associated with the increasing use of steep lands and semi-arid lands are discussed in later chapters.

The fact of soil erosion being a serious problem to sustained food production has long been recognized. The extent of erosion damage was very clearly stated more than 50 years ago by JACKS and WHYTE (1939) and again by VOGT (1948), but quantitative assessments are few and far between. One of the earliest and most reliable, and which had a dramatic impact, was that carried out in the United States of America in 1934 which showed that almost three quarters of the land under cultivation had been seriously damaged (BENNETT 1939). Subsequent assessments in the USA and elsewhere have confirmed that this is not an isolated example, and recently we have come to realize that assessments of the tons of soil lost are not efficient measures of the loss of soil productivity which is what we really need to know.

In recent years many countries have used our increasing understanding of the process of soil erosion to assess and map the relative risk of erosion in different places, and in southern Africa this has now been completed for most of the ten countries which comprise the Southern African Development Community (SADC), using a methodology developed by STOCKING (1987).

At the international conference on the human environment in Stockholm 1972 and conferences on degradation organized since then by FAO, UNEP and ISSS, a plan was agreed for a major international cooperative study under the title Global Assessment of Soil Degradation (GLASOD) which resulted in the publication of a world map of the status of human-induced soil degradation at a scale of 1:10 m. on three sheets (OLDEMAN *et al* 1991). Over 250 soil and environmental scientists operated in this programme and the results are stored in a digital database so that additional information and modifications can be added at a later date and the map can be reproduced at other scales to give more detailed information for a country or region. Some of the results are summarized in tables 1.2 and 1.3.

1.2 The possibilities of increasing food production

Food production is increasing and, on a world scale, increasing faster than population (table 1.4) but the global figure is irrelevant because the food deficiency in some continents cannot be met by transferring food from those countries with a surplus. The inter-continental movement of food is important in terms of famine relief and international trading, but there simply are not sufficient trains and ships to move enough food to overcome the shortages which now exist and are getting worse. The danger spots are the countries which show a steady decline in food production per head, mainly in Africa (figure 1.3).

1.2.1 Increasing area

Attempts to assess the amount of potentially arable land which could be brought into cultivation face several difficulties. FAO has used its huge database of world soils and climate to calculate that the presently cultivated 1.5 billion hectares could theoretically be almost doubled to 2.8

Table 1.2 Human-induced soil degradation for the world (millions of hectares)

	*Light**	*Moderate*	*Strong*	*Extreme*	*Total*	
Water	343	527	217	7	1094	(56%)
Wind	269	254	24	2	549	(28%)
Chemical	93	103	42	1	239	(12%)
Physical	44	27	12	—	83	(4%)
Total	749 (38%)	911 (46%)	295 (15%)	10 (0.5%)	1965	

* Light degradation implies somewhat reduced productivity, manageable in local farming systems.
Moderate degradation has greatly reduced productivity, restoration requires improvements beyond the means of local farmers in developing countries.
Strong degradation is soils not reclaimable at farm level, restoration requires major engineering work or international assistance.
Extremely degraded soils are beyond restoration.
Data summarised from OLDEMAN *et al* 1991

Table 1.3 Causative factors of soil degradation (millions of hectares)

	Deforestation	Overgrazing	Agricultural Mismanagement	Over-Exploitation
Africa	67	243	121	63
Asia	298	197	204	46
S. America	100	68	64	12
N.+C. America	18	38	91	11
Europe	84	50	64	1
Australasia	12	83	8	—
WORLD	579	679	552	133

Data summarised from OLDEMAN *et al* 1991

billion hectares (FAO 1992). But there are constraints on this happening – much of the potential land would require drainage, or irrigation, or the correction of chemical problems, involving large development costs; draining wetlands or clearing forests may be ecologically undesirable; some areas may be unusable until health hazards are overcome; in many countries land development is held up, or even thrown into reverse, by civil unrest; and the most important limitation is the distribution of

Table 1.4 Trends in food production per head

	Index of food production per person (1979–81 = 100)		
	1964–66	*1979–81*	*1984–86*
World	94	100	104
Africa	108	100	95
14 Countries show fall in both periods			
eg Angola	127	100	90
Botswana	134	100	79
Kenya	119	100	85
4 Countries show rise in both periods			
Libya	78	100	133
Tunisia	96	100	107
Benin	94	100	114
Cote d'Ivoire	73	100	101
North & Central America	90	100	99
South America	91	100	101
Asia	92	100	112
2 Countries show a fall in both periods			
Bangladesh	119	100	99
Mongolia	136	100	96
15 Countries show a rise in both periods			
eg China	80	100	122
India	93	100	111
Europe	82	100	107
USSR	86	100	107
Oceania	92	100	99

World Resources Institute 1988, chapter 17

potentially arable land (table 1.5). In seven countries in South Asia 90% of the usable land is already in use and on average supports 8.5 people per cultivated hectare. At the other end of the scale, in South America, only 15% is currently used and supports 2 people per hectare (FAO 1992).

1.2.2 Increasing yields

Here is the greatest opportunity. The average yield of cereals in many parts of Africa is below one tonne of grain per hectare, while experimental stations in Africa consistently demonstrate that yields twice or three times greater can be achieved by applying simple methods, and intensive production can yield 8 tonnes per hectare. The country average in the USA is 4.3 tonnes per hectare (WORLD RESOURCES INSTITUTE 1992–93, table 18.1). Similarly, much of the world's rice is produced at yields

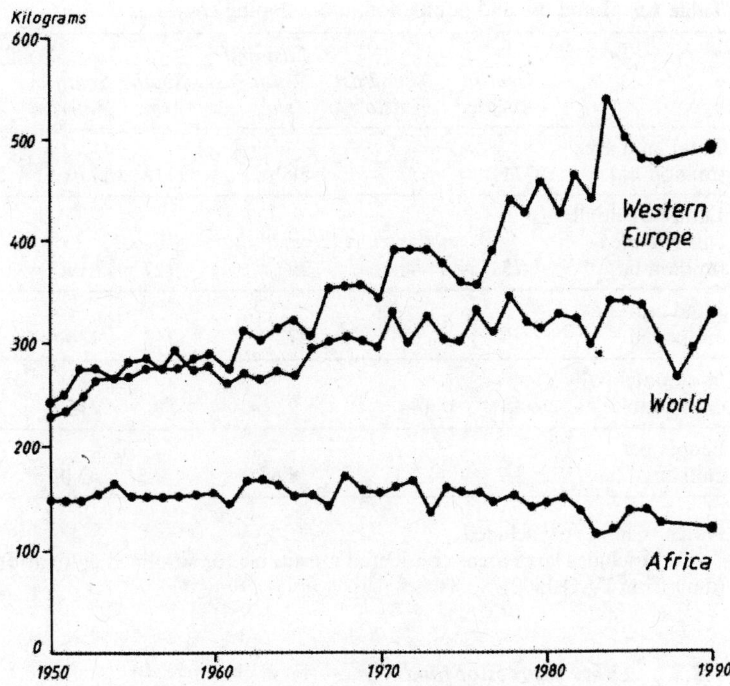

Figure 1.3 Annual grain production per person in the World, Western Europe, and Africa (WORLDWATCH INSTITUTE 1988)

only a fifth or a quarter of that which could be obtained. In order t achieve increased yields, the primary requirement is not research into new methods, but the increased application of techniques and practices which are already known.

To quote a recent analysis of this problem: 'Land with potential for crops is not available to feed the doubled population at the present very low crop yields. Higher yields are essential for survival. The new strains of disease-resistant and fertilizer-responsive crops offer the only known solution to meeting so vast an increase in demand. Many high-yielding varieties are available. The most critical constraints are the supply and distribution of fertilizers and the organization of credit systems to enable farmers to use them' (PEREIRA 1993). Another problem is that the organic matter content of tropical soils is often depleted by the shortage of domestic fuel. The burning of dung and crop stalks must be countered by the planting of fast-growing fuelwood trees along boundaries and roadsides.

Table 1.5 Land use and population in developing countries

	Central America	*Near East in Asia*	*East and South-east Asia*[1]	*South Asia*	*South America*	*Africa*
Total land area (million ha)	272	677	897	1116	1770	2886
Land potentially cultivatable (million ha)	75	48	297	127	819	789
Land currently cultivated	36	69[2]	272	113	124	168
% of potentially cultivatable	48	144	92	89	15	21
People per cultivated ha	3.2	2.1	4.6	8.5	2.0	2.5

Notes [1] China not included
[2] Includes large areas considered unsuitable for sustained cultivation
(data from FAO 1992)

1.2.3 Other sources of food

Man is a resourceful creature, particularly when his survival is threatened, and there is a strong possibility that hitherto untapped sources of food will be sought out and developed. The chemical food pill of science-fiction writers is still impractical, but three possible sources, all being seriously investigated today, are the large-scale exploitation of fish foods, the processing of lower forms of marine life, such as algae and plankton, and the manufacture of synthetic protein.

Fish, when produced in artificial ponds as a crop which is established, grown, and harvested, can give a higher yield in terms of kilograms per hectare than an irrigated vegetable crop, and further provide the animal protein which is the main deficiency in the diet of most underfed people. In natural streams, rivers, and lakes, 'fish farming' cannot be as precise as in artificial ponds, and the yields are correspondingly lower, but the inland waters of the world are producing an increasing proportion of the global fish catch, from 2% in 1976, to 7% in 1991, with another 5% from marine aquaculture (FAO 1991).

The fish yield from the world's oceans is thought to be, in global terms, close to the estimated sustainable yield of about 95–100 million tonnes, but in many of the main fishing areas long-term productivity is endangered by overfishing, in the same way that the soil resources can be damaged by excessive exploitation (WORLDWATCH INSTITUTE 1993 page 8).

The idea of algae or plankton as a human food is not attractive at present, but the growth of such forms of life is a highly efficient way of converting chemical nutrients and the sun's energy into edible substances, and the product might be made palatable in either of two ways; by treatment, including the addition of synthetic tastes, so that the origin was not recognizable, or by using it as a high-quality feed for conventional food like pigs, poultry, or cattle.

The manufacture of synthetic protein is at an early stage of development, but biotechnology is one of today's growth industries and rapid progress is likely. The two main approaches are single cell micro-organisms converting petroleum products, and yeast fermentation of inedible crop residues such as straw. It is human nature to be very resistant to change in dietary habit and taste, so it is probable that, like algae, the product will be used as a feed for livestock or fish rather than for direct human consumption.

The present position appears to be that there are limited opportunities for expanding conventional food supplies, and unproved opportunities for developing new supplies, but the enormous gulf between supply and demand focuses attention on the vital need to prevent any running down of the soil's productivity.

1.3 The historical background of soil erosion

Today soil erosion is almost universally recognized as a serious threat to man's well-being, if not to his very existence, and this is shown by the fact that most governments outside Europe give active support to programmes of soil conservation. But it is relevant, before making any assessment of present knowledge of erosion, to consider the development of this science which was almost unknown 80 years ago, and now enjoys world-wide attention.

Studies of the effect of erosion on early civilizations have shown that a major cause of the downfall of many flourishing empires was soil degradation (LOWDERMILK 1953). Although this is clearly evident throughout 7 000 years of history, an awareness of the problem developed very slowly.

There are passing references in the Old Testament, mainly threats of streams drying up. Occasionally Greek writers mention the problem, eg Homer on fallows, Plato on floods and deforestation. The Romans had a better understanding, with Virgil advocating what today we would call conservation farming, but the main point is that it was not until the beginning of the present century than an appreciation of the problem was sufficiently widespread for governments to start taking an interest.

Part of this reluctance to appreciate the significance of erosion may stem from the fact that the earliest civilizations all arose on irrigated alluvial plains, and many were dependent upon flood deposits of silt for continued fertility. The civilizations of the valleys of the Nile, the Tigris, and the Euphrates, which owed their existence to erosion in the headwaters, could hardly be expected to see erosion in the same light as a modern agricultural community.

There is some debate about whether long-term changes in climate have affected past soil degradation. The conventional view was that no climatic change has occurred (REIFENBERG 1955), but then VITA-FINZI (1969) showed that major geological changes have taken place in the Mediterranean in historical times, and PARRY (1978) suggests that climatic changes may have been more important than was previously assumed. However, the evidence of protected temple forests in both China and the Lebanon suggests that whether or not climate has been a contributing factor, the devastation we see today is essentially a man-made phenomenon (LOWDERMILK 1953).

1.4 The growth of erosion research

The first scientific investigations of erosion were carried out by the German soil scientist WOLLNY, between 1877 and 1895 (BAVER 1939). Small plots were used to measure a wide range of effects, such as that of vegetation and surface mulches on the interception of rainfall and on the deterioration of soil structure, and the effects of soil type and slope on runoff and erosion. Apart from this pioneer work, the lead in erosion research has come mainly from the United States of America. Isolated cases of practical application by farmers of mechanical conservation works increased from the 1850s until, in 1907, the United States Department of Agriculture declared an official policy of land protection.

The first American quantitative experiments were laid down by the Forest Service in 1915 in Utah, closely followed by those of Miller in Missouri in 1917, which led in 1923 to the first published results of field plot experiments. Other similar experiments followed, using essentially the same method, and were given added impetus by the allocation of funds by Congress in 1928. These enabled Bennett to establish between 1928 and 1933 a network of ten field experiment stations. During the next decade this programme expanded until 44 stations were operating, and included experiments on mechanical erosion control and runoff from small catchments.

Throughout this period the work was limited to applied research, in which problems were studied under field conditions; and, although it had been apparent from the earliest days of Wollny's work, that the prevention of splash erosion was of vital importance, there was no coordinated research involving an analytical study of the process of erosion. Pioneer work in this field was carried out in the 1930s by a few individuals such as Baver, Borst, Woodburn, and Musgrave, and led to the first detailed study of natural rain by LAWS in 1941, and the first analysis of the mechanical action of raindrops on the soil by ELLISON in 1944. The implications of this are best described by STALLINGS (1957 chapter 1) who says:

> 'The discovery that raindrop splash is a major factor in the water erosion process marks the end of one era in man's struggle with soil erosion and ushers in another which, for the first time, holds out hope for a successful solution to the problem. The exact nature of the effects of raindrop splash is the phase of the water-erosion process that escaped detection during the first 7000 years of civilization. It explains why the efforts at protecting the land against scour erosion these 7000 years have failed. It explains why there is little or no erosion on land with ample plant cover. It explains many things that have puzzled agricultural leaders and practitioners throughout this long and troublesome period.'

It remained for Ellison to recognize the true role of the falling raindrop in the water erosion process. He was the first to realize that the falling raindrop was a complete erosive agent within itself and that little or no erosion occurred when the ground surface was protected by ample cover. He showed that the protective effect of plant cover was due to the fact that it robbed the falling raindrop of its kinetic energy. Ellison's discovery opened a new field of soil erosion science.

Analytical research was directed to more specific objectives by the setting up in 1954 of a national study, which used modern techniques of data analysis to correlate the results of all the field experiments (WISCHMEIER 1955). As a result of this study, the main features in the erosion process were identified and mathematically enumerated (WISCHMEIER *et al* 1958). This work ushered in the phase of quantitative scientific investigation discussed in subsequent chapters. A major recent change has been the shift in emphasis from 'black-box' empirical models such as USLE to process-based models such as WEPP and GUESS, discussed in chapter 7.

The United States has always maintained a commanding lead, both in research on the erosion process and in studies of the application of conservation practices. Many other countries have followed with national programmes designed to test the relevance of USA work in other conditions, especially in tropical countries where erosion is particularly severe. The search for new conservation methods has been stimulated by the belated realization that the protection methods effective in the United States mid-West cannot always be extrapolated to other situations, for example small-scale subsistence agriculture on steep lands in the tropics. So the new thrust of erosion research in developing countries is a search for new locally-applicable solutions instead of an attempt to adapt or modify imported methods. Another important change has been to give much higher priority to erosion control through better farming and less attention to mechanical protection works (SHAXSON *et al* 1989). National programmes have been reinforced by international programmes. The international agencies such as FAO, UNEP and UNESCO have become involved in action programmes, and the international agricultural research stations, particularly ICRISAT in India, and ICRAF in Kenya, have started multi-national research on erosion.

The opportunities for international exchange of new ideas and research methods have increased out of all recognition in the last ten years. The main forum is the ISCO series of biennial conferences, supported by *ad hoc* workshops and regional conferences which are listed in *Further reading*.

1.5 The geographical distribution of erosion

The two main agents of erosion are wind and water, and by considering the conditions under which each will be active, a pattern can be built up of the areas of the world where either wind erosion or water erosion is likely to be particularly serious.

1.5.1 Erosion by water

The factor which most influences soil erosion by water is the mean annual rainfall, as shown in figure 1.4. In regions of very low rainfall there can naturally be little erosion caused by rain. Further, what little rain does fall is mainly taken up by a vegetation permanently short of water so there is little runoff. At the other extreme, an annual rainfall of more than 1 000 mm usually leads to dense forest vegetation. This affords protective cover to the soil and, as is shown in the next chapter, the presence of cover is the key factor in reducing water erosion. The most severe erosion will thus tend to be associated with the middle range of rainfall when the natural vegetation is largely undisturbed, and with higher rainfall when the natural forest is removed. A summary of world rainfall distribution is shown in figure 1.5

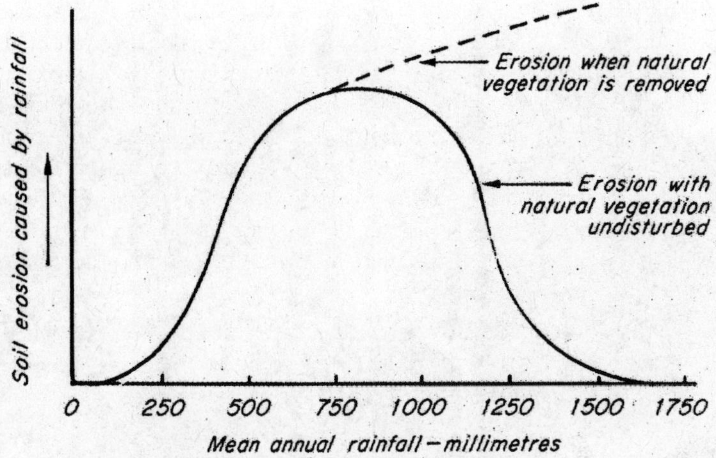

Figure 1.4 The relation between rainfall and soil erosion

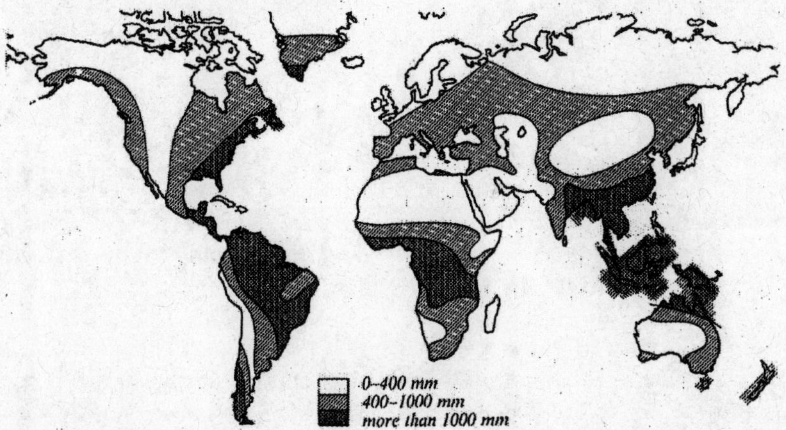

Figure 1.5 Generalized map of mean annual rainfall

However, it is not only the amount of rainfall that matters, but also the kind of rain. The intensive downpour common in the tropics has a much more damaging effect than the gentler rain of temperate climates, and the approximate limits of the area of destructive rain are latitudes 40° North and 40° South. There are, of course, exceptions to this world-scale pattern. In semi-arid conditions, serious rain erosion often occurs because the rain, although low in quantity, comes in very severe storms (plate 1.1). In other

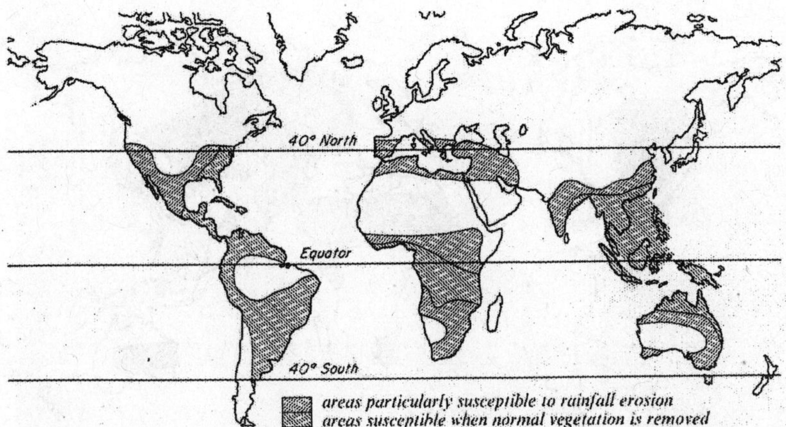

Figure 1.6 Generalized map of the geographical distribution of rainfall erosion

cases, steep slopes and vulnerable soils can lead to quite serious erosion in temperate latitudes. In general, however, soil erosion by water can be expected to be most serious in areas between these latitudes, where the annual rainfall is neither very high nor very low. Figure 1.6 shows that, as a first approximation, this is in fact the case, the main areas being North America up to about 40° N, parts of South America, nearly all of Africa except the dry deserts and equatorial forest, Asia up to 40° N, and Australia excluding the dry centre.

1.5.2 Erosion by wind

There are also two main conditions which must exist before wind erosion can be a serious problem. Firstly, only dry soil blows, so the vulnerable regions are those with a low mean annual rainfall, particularly less than about 250 to 300 mm. Secondly, large-scale movements can occur only where there are steady prevailing winds at all levels from the upper air down to ground level, and these are associated with large, fairly level land masses. Naturally, there exist local exceptions, but on a continental scale, wind erosion is found to be most serious where it would be expected from consideration of these two requirements. The main areas, shown in figure 1.7, are North America (the Great Plains, famous as

Plate 1.1 Rainfall erosion is greatest when heavy rainstorms fall on land unprotected by vegetation (Copyright Aero Films Limited)

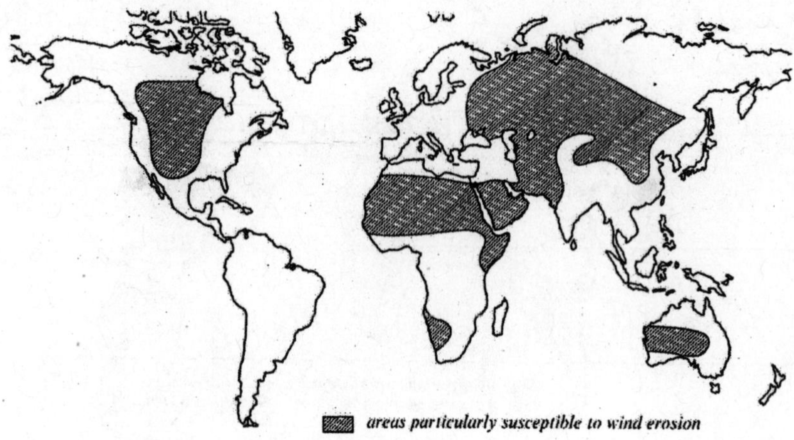

Figure 1.7 Generalized map of wind erosion

the Dust Bowl), the Sahara and the Kalahari deserts in Africa, central Asia (particularly the steppes of Russia), central Australia, and central China.

1.5.3 The extent of soil degradation

Wind and water erosion cannot occur at the same time, because soil only blows when it is dry. They can occur at the same site at different times. In terms of geological erosion deposits of fine material from water erosion can later be vulnerable to wind erosion in a subsequent drier climate as described by MCTAINSH *et al* (1992). On a shorter time-scale, cultivated land may be subject to wind erosion during a dry windy season (common spring or autumn), and to water erosion during a season of heavy rainfall (usually summer). But both these occurrences are infrequent and unimportant. The usual situation is that erosion is either by wind or water. Globally more land is damaged by water erosion than by wind erosion (table 1.2), but the relative proportions vary considerably in different continents reflecting the patterns shown in figures 1.6 and 1.7. The extent of land degradation on each continent has been computed as part of the GLASOD project (OLDEMAN *et al* 1991). However, all estimates of degradation at regional or continental scale should be treated with caution. LAL (1988) has shown that the apparent results may be reflecting the assumptions made and the methods of measurement and analysis rather than real differences from place to place. Figure 1.8 shows unacceptable variation between different studies of the same parameter – the suspended sediment yield in streams and rivers.

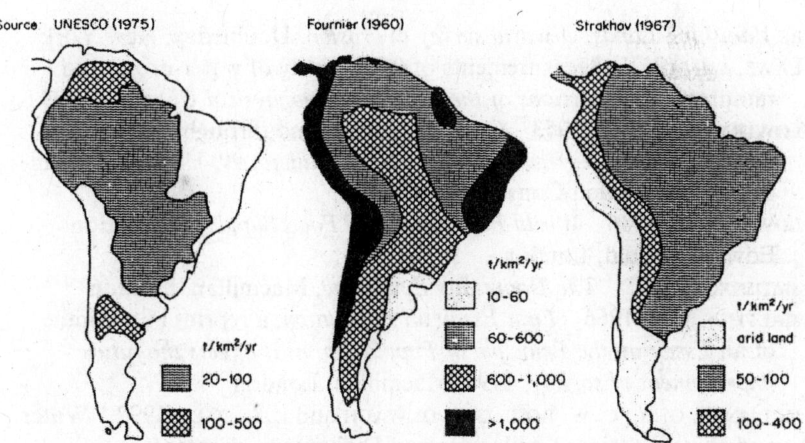

Figure 1.8 Varying estimates of suspended sediment production.
Sources:
STOCKING, M. A. (1993) Soil Erosion in Developing Countries. Discussion Paper 241, School of Development Studies, University of East Anglia.
LAL, R. (1988) Soil Erosion by wind and water. Problems and prospects. In: Soil Erosion Research Methods, R. Lal (ed), Soil and Water Conservation Society, Ankeny, Iowa.
UNESCO (1975) Gross sediment transport into the oceans. UNESCO/IAHS. Paris.
FOURNIER, F. (1960) Climat et Erosion. Presses Universitaires de France, Paris.
STRAKHOV, N. M. (1967) Principles of lithogenesis (volume 1). Oliver and Boyd, Edinburgh.

References

BAVER, L. D. 1939 Ewald Wollny, A pioneer in soil and water conservation research, *Soil Science Society of America*, 3, 330–333

BENNETT, H. H. 1939 *Soil Conservation*, McGraw–Hill, New York

BOERMA, A. H. 1975 The World Could be Fed, *Journal of Soil and Water Conservation*, 30, 1, 4–10

BOSERUP, E. 1993 *The Conditions of Agricultural Growth*, Earthscan, London (first published by Allen & Unwin, London 1965)

ELLISON, W. D. 1944 Studies of Raindrop Erosion, *Agricultural Engineering*, 25, 131–136, 181–182

FAO 1991 *Environment and Sustainability in Fisheries*, FAO, Rome

FAO 1992 *Protect and Produce*, 2nd edition, FAO, Rome

JACKS, G. V. and R. O. WHYTE 1939 *The Rape of the Earth: A world survey of soil erosion*, Faber and Faber, London. Published in USA as *Vanishing Lands: A world survey of erosion*, Doubleday, New York

LAWS, J. O. 1941 Measurements of fall-velocity of water-drops and raindrops, *Transactions of the American Geophysical Union*, 22, 709

LOWDERMILK, W. C. 1953 Conquest of the land through seven thousand years, *Agriculture Information Bulletin 99*, US Department of Agriculture Soil Conservation Service

LOWRY, J. H. 1986 *World Population and Food Supply*, 3rd edition, Edward Arnold, London

MADDOX, J. 1972 *The Doomsday Syndrome*, Macmillan, London

MALTHUS, T. R. 1966 *First Essay on Population*, a reprint in facsimile of an *Essay on the Principle of Population, as it affects the future improvement of society*, 1798, Macmillan, London

MCTAINSH, G. H., C. W. ROSE, G. E. OKWACH and R. G. PALIS 1992 Water and Wind Erosion: Similarities and Differences, chapter 9, in *Erosion, Conservation, and small-scale Farming*, edited by H. Hurni and K. Tato, *Geographica Bernensia*

MEADOWS, D. H., D. L. MEADOWS, J. RANDERS and W. W. BEHRENS III, 1972 *The Limits to Growth*, Earth Island, London

MEADOWS, D. H., D. L. MEADOWS and J. RANDERS 1992 *Beyond the Limits*, Earthscan Publications, London

OHADIKE, P. O. 1992 Population, Policy Development and Implementation in Sub-Saharan Africa, *African Development Review*, 4, 2, 273–297

OLDEMAN, L. R., R. T. A. HAKKELING and W. G. SOMBROEK 1991 *World Map of the Status of Human-Induced Soil Degradation*, International Soil Reference and Information Centre, Wageningen

PARRY, M. L. 1978 *Climatic Change, Agriculture and Settlement*, Dawson & Sons, Folkestone, Kent

PEREIRA, SIR CHARLES H. 1993 Food production and population growth, *Land Use Policy*, July 1993, 187–190

REIFENBERG, A. 1955 *The Struggle between the Desert and the Sown; Rise and Fall of Agriculture in the Levant*, The Jewish Agency, Jerusalem, Israel

REVELLE, R. 1976 The Resources Available for Agriculture, *Scientific American*, 235, 1, 164–178

SHAXSON, T. F., N. W. HUDSON, D. W. SANDERS, E. ROOSE and W. C. MOLDENHAUER 1989 *Land Husbandry: A Framework for Soil and Water Conservation*, Soil and Water Conservation Society, Ankeny, Iowa

STALLINGS, J. H. 1957 *Soil Conservation*, Prentice-Hall, Englewood Cliffs, New Jersey

STOCKING, M. A. 1987 *A Methodology for Erosion Hazard Mapping of the SADCC Region*, Report 9, SADCC-ELMS, Lesotho

TIFFEN, M., M. MORTIMER and F. GICHUKI 1993 *More People, Less Erosion, Environmental Recovery in Kenya*, Wiley, Chichester, Sussex

VAIDYANATHAN, K. E. 1992 Population Trends, Issues and Implications, *African Development Review*, 4, 2, 1–32

VITA-FINZI, C. 1969 *The Mediterranean Valleys – Geological Changes in Historical Times*, Cambridge University Press

VOGT, W. 1948 *Road to Survival*, Sloane, New York

WISCHMEIER, W. H. 1955 Punched cards record runoff and soil loss data, *Agricultural Engineering*, 36, 664–666

WISCHMEIER, W. H., D. D. SMITH and R. E. UHLAND 1958 Evaluation of Factors in the Soil-Loss Equation, *Agricultural Engineering*, 39, 8, 462–474

WORLDWATCH INSTITUTE 1988 The changing world food project: the nineties and beyond, *Worldwatch Paper 85*, Worldwatch Institute, Washington DC

WORLDWATCH INSTITUTE 1993 *State of the World 1993*, Earthscan Publications, London

WORLD RESOURCES INSTITUTE 1988 *World Resources 1988–89*, Basic Books, New York

WORLD RESOURCES INSTITUTE 1992 *World Resources 1992–93*, World Resources Institute, Washington DC

2

The mechanics of erosion

2.1 Geological and accelerated erosion

Erosion has always taken place, and always will. The surface of the earth is constantly changing, with mountains rising, valleys being cut deeper and wider, the coast line receding here and advancing there. The physical pattern of the surface of the earth which we see today is not the result of some single cataclysmic sculpturing but the result of changes so infinitely slow that only after centuries is the effect noticeable. Erosion is simply one of the aspects of the constant process of change. It is fundamental to the formation of alluvial soils and sedimentary rocks. Man's activities seldom slow down or halt the process but frequently speed it up. We usually refer to *geological erosion* or *normal erosion*, or *natural erosion* when we mean that which results only from the forces of nature, and to *accelerated erosion* when the process is influenced by man.

When trying to predict the probable severity of man-made erosion it helps to consider the rate of geological erosion. If the conditions of climate and topography are such that geological erosion is quicker than usual, then these same conditions will also lead to particularly severe accelerated erosion. Extreme examples of both kinds of erosion are shown in plates 2.1 and 2.2.

2.1.1 The agents of erosion

Erosion is essentially a smoothing or levelling process, with soil and rock particles being carried, rolled, or washed down by the force of

Plate 2.1 Geological erosion can be spectacular. This erosion is the result of highly erosive rain falling on soil with low resistance to erosion and with no vegetative cover (Copyright Fairchild Surveys)

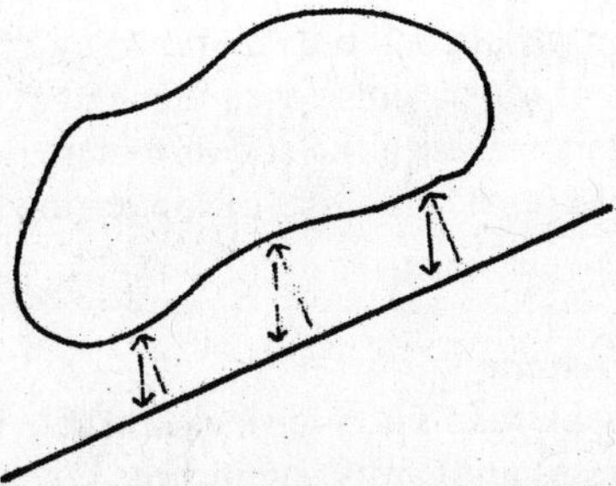

Figure 2.1 Frost action. The expansion caused by freezing is at right angles to the soil surface. On thawing the particle sinks vertically. There is a net movement downslope

gravity. The main agents which loosen and break down the particles are wind and water.

Wind does not by itself wear away rocks, but abrasion, even of hard rock, results from grains of sand or soil carried in suspension – rather like a slow-motion version of the process of sand-blasting used to clean metal surfaces before painting.

Water is probably the most important single agent of erosion. Rainfall, streams, and rivers all scour away or carry away soil. Waves erode the shores of seas and lakes – in fact wherever water is in movement it is eroding its boundaries.

Temperature changes When considering geological erosion the passage of time is hardly noticeable, and even minutely small or incredibly slow changes become significant over a long period of time. Examples are the cracking and flaking of rocks by variations in temperature. Rapid variations between day and night affect only the surface of rocks while the changes due to slower variations between summer and winter penetrate deeper. When temperature changes include frost, the disruption is greatly increased by the expansion of water in cracks and crevices. Frost heave and traction results in a progressive movement downhill in sloping ground as shown in figure 2.1.

Biological Some actual destruction may be caused by living organisms, such as lichens and mosses on rocks, but the main effect of living things is the disturbance which speeds up the effect of other agents.

Plate 2.2 This field in Tanzania lost 50 mm of topsoil over its whole surface with rills cut down 150 mm to the plough depth, during a storm of only a few hours (Copyright National Institute of Agricultural Engineering)

Animals trampling on rock or soil break it down and make it more easily carried away by wind or water, and at the other end of the biological scale earthworms and termites disturb the soil and increase the aeration and oxidation, and so speed up the processes of conversion from resistant rock to erodible soil.

2.1.2 Accelerated erosion

On a world scale man's non-agricultural activities which accelerate the erosion processes are hardly significant. We excavate mountains for coal and ores, we dig a hole here and fill in there, we clear vegetation and build on the site, but all this interferes with only a small part of the earth's surface. On the other hand, agriculture is so widespread that agricultural activities which materially alter the speed of the erosion process are much more important. And nearly all agricultural operations do tend to encourage erosion. Whenever vegetation is cleared and the ground is more exposed, there are fewer trees to slow down the wind, so wind erosion increases, less vegetation to absorb the energy of the falling rain so more rainfall erosion, more surface runoff so streams and rivers become stronger, more cattle to crush the rock and soil. By ploughing and tilling the soil man disturbs and aerates the soil millions of times more quickly and effectively than the burrowing animals – in fact all the physical processes of nature are accelerated, and so is erosion. Only in isolated cases is there a possibility of natural erosion being reduced by man where desert areas are reclaimed, or arid regions made temperate by irrigation, or forests established, and these are infinitesimal compared with the areas where it is increased.

2.1.3 Acceptable limits of erosion

What then is the dividing line between geological erosion, a natural phenomenon we must accept, and accelerated erosion, a man-made destructive thing? Since it will be difficult, if not impossible, to draw this line, is another possibility to specify the limits of erosion which can be accepted as tolerable? The question then becomes: 'What is the level of erosion at which we change from accepting it to feeling that something ought to be done about it?' The usual answer is that the object of soil conservation is to ensure that land is used in such a way that the use can be sustained indefinitely, i.e. there is no progressive deterioration. It is sometimes assumed that the criterion should be when soil loss is no greater than the rate of formation of soil, but this does not help because the rate of formation cannot be precisely measured. The best estimate of soil scientists is that under undisturbed conditions it will take of the order of

300 to 1000 years to form 25 mm of soil but that when the disturbances and aeration and leaching actions are speeded up by tilling the land, this will be reduced to something like 100 years (PIMENTEL *et al* 1976). Some of the many other estimates of rates of soil formation are reviewed by MORGAN (1986 page 162). Also, equating acceptable soil loss simply with formation of new soil takes no account of possible improvements to soil productivity or fertility by good management.

The level of erosion which has for many years been taken in the United States of America as the upper limit of tolerable soil loss is 11.2 tonnes/hectare/year equivalent to a formation rate of 25 mm in 30 years. There is no need for the apparent precision of the number 11.2. It is only the metric conversion of the original unit of 5 tonnes/acre/year which was the round number agreed by panels of experts in six regions of the USA in the 1960s. It is sometimes stated that the values were selected for reasons such as that greater losses would adversely affect water control structures, that higher losses would be accompanied by gully formation, and that the cost of replacing nutrients would be uneconomic at higher levels. In fact these considerations have been added later in justification for retaining the originally chosen values. If we go back to the reports of the six regional workshops we find that the reason was: 'We consider these to be rates that will make it possible for the operator to maintain the soil in a productive state through a long period of years and without serious sedimentation and without having great losses of plant nutrients' (EDWARDS 1961). It is also sometimes mistakenly assumed that the 5 tonnes/acre/year was a maximum figure to be applied generally to most soils. In fact, in the report of the regional workshop for Arkansas, Louisiana, Mississippi, Oklahoma and Texas just qoted we find that of the 51 soil types allocated a *T* value, more than three quarters had values between 2 and 4. All soils in the USA have been assigned *T* values ranging from 1 to 5 tonnes per acre per year (4.5 to 11.2t/ha/y), and a number of other soil conservation services have adopted similar values. Naturally any value of acceptable loss will depend on the soil conditions – if the profile consists of a deep soil whose fertility is the same at all depths, then to lose 25 mm of soil in 30 years is much less serious than if the profile consists of a few centimetres of soil overlying hard rock.

The validity of using values chosen without firm evidence is frequently challenged and was discussed at the 1979 annual meeting of the Soil Science Society of America. While many doubts were expressed about the adopted values, there was no move to substitute others (MANNERING 1981). Scientists sometimes worry about the lack of experimental evidence supporting the chosen values (JOHNSON 1987), overlooking the fact that *T* values are not intended to represent directly the physical processes of soil formation, and are merely a tool to assist in the process of planning and choosing alternative forms of land use (as discussed in section 7.2.2). Recently in the United States of America there have been moves to coerce landowners into adopting improved farming practices, discussed in section 14.2.4. Acceptable practices are defined as whose which according to the Universal Soil Loss Equation (USLE) will result in an average loss less than a stated value. Initially the values were set at twice the SCS chosen *T* values, but the farmers complained, demanding that these limits should be doubled to $4 \times T$, and they were supported by the head of the service in Washington. Next the SCS staff rebelled at this slackening of the standard, and the legislation is, for the moment, back to $2 \times T$, but the argument continues (PADGITT and LASLEY 1993). Two conclusions can be drawn from this experience. Firstly, the SCS has by suggesting $2 \times T$ admitted truly sustainable farming with no degradation is not an achievable target, since this would require using a limit of *T*, and the farmers demanding $4 \times T$ are showing that they are willing to accept an even higher rate of degradation. Secondly, there can be little hope of enforcing through the courts, legislation which is based on converting arbitrarily chosen targets into legal binding requirements. The relation between acceptable limits and farming practice is further discussed in section 7.2.2 and the question of legislation in section 14.2.4.

2.2 Forms of erosion

2.2.1 Water erosion

The first classification of water erosion attempted by early conservationists was into stages corresponding with the progressive concentration of surface runoff. It started with sheet erosion (the washing of surface soil from arable lands), then rill erosion as the water concentrates into small rivulets in the fields, then gully erosion when the eroded channels are larger, and finally streambank erosion when rivers or streams are cutting into the banks. This classification is no longer appropriate to our understanding of the erosion process, and can be misleading for it omits the splash or impact effect of the raindrops which we now know to be the first and most important stage in the erosion process. Also, sheet erosion conjures up a picture of soil being removed uniformly in thin sheets by the even flow of water. This is wrong on all accounts. Runoff seldom occurs as smooth laminar flow, and in any case smooth laminar flow only scours soil at velocities much higher than are usually found in runoff. A better approach is to describe this first phase as *inter-rill* erosion, meaning both movement by rain splash and the transport of raindrop-detached soil by thin surface flow whose erosive capacity is increased by turbulence generated by raindrop impact.

If we discard the description 'sheet-erosion', and substitute 'inter-rill erosion' there is no objection to continuing the sequence: rill, gully, and streambank. *Rill erosion* is usually defined as small washes which can be eliminated by normal cultural methods, and *gullies* when the channels are so large and well-established that they cannot be crossed by farm implements. There is no precise dividing line between the two.

More detailed accounts of the processes of erosion are listed in *Further reading*.

2.2.2 *Specialized forms of erosion*

Pedestal erosion (plate 2.3)

When an easily eroded soil is protected from splash erosion by a stone or tree root, isolated 'pedestals' capped by the resistant material are left standing up from the surrounding ground. The erosion of the surrounding soil is shown to be mainly by splash rather than surface flow because there is little or no undercutting at the base of the pedestal. This type of erosion develops slowly over several years and is often found on bare patches of grazing land. It can occur in arable lands which suffer excessive erosion during exceptional storms. The main interest is that it is possible to deduce approximately what depth of soil has been eroded by studying the height of the pedestals.

It is important to differentiate between pedestals and stools of grass which frequently have a soil level above the surrounding ground. Such raised soil levels may show the original level with the soil between the grass clumps having been eroded, but it is more likely that the level in the grass clumps has been raised by catching soil splashed from the surrounding bare patches. On experimental plots the soil level in tufts of *Eragrostis curvula* (weeping love grass) was 20 mm above the level between the tufts, and it might easily be assumed that this is evidence of considerable erosion. In fact measurements showed that the soil loss was negligible and this effect was solely due to the lateral movement of the soil by splash.

Pinnacle erosion

The characteristic erosion pattern which leaves high pinnacles in gully sides and bottoms is usually associated with 'difficult' soils which are highly erodible (plate 2.4). This erosion is always associated with deep vertical rills in the gully sides, and these cut back rapidly until they join and leave the isolated pinnacles. A more resistant soil layer, of gravel or

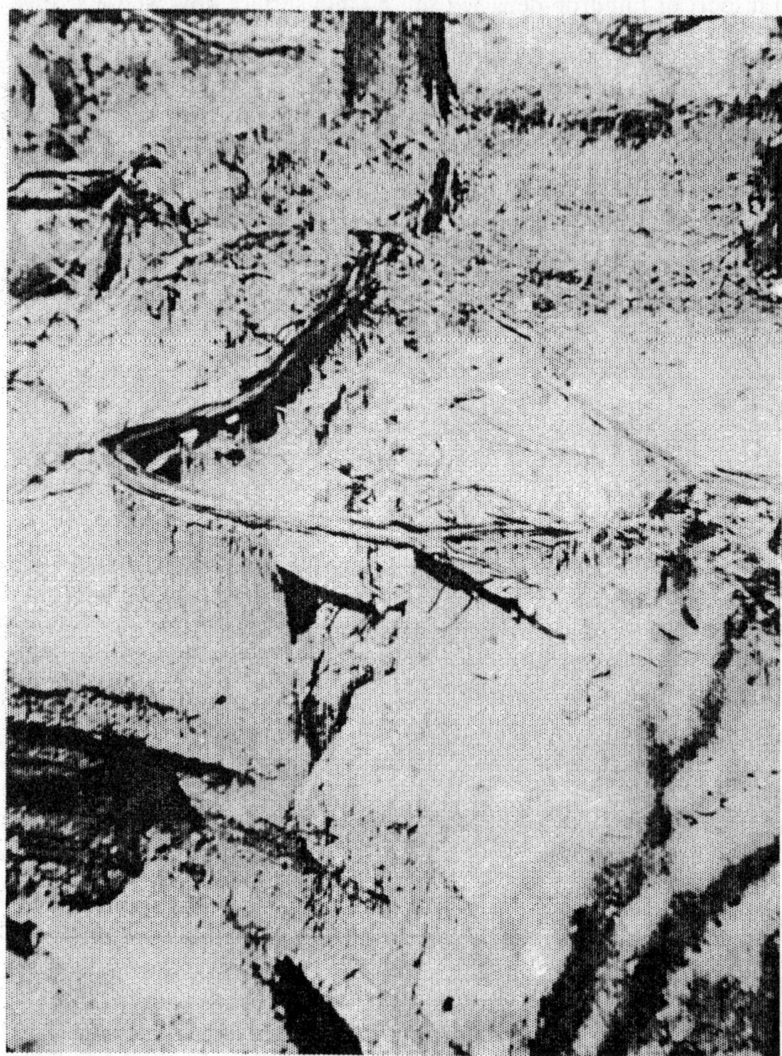

Plate 2.3 Pedestal erosion results when a stone or tree root protects the soil from the splash erosion which removes the surrounding soil

Plate 2.4 Pinnacle erosion often occurs in gullies as the result of deep vertical rills widening until pinnacles are left like islands in the bed of the gully

stones, often caps the pinnacle as in pedestal erosion. Banks eroded in this form are usually severely undercut by either flowing or standing water, and pipe erosion is also frequent. The chemical or physical conditions which cause this form of erosion are not clearly defined, but it is usually found where there is some severe imbalance such as excessive sodium and complete deflocculation. Soils liable to this type of erosion are recognized by the fact that when dry they take up water very slowly and reluctantly but, once saturated, they have no cohesion and flow like mud.

Gully control or any reclamation is always difficult in soils showing pinnacle erosion. Adverse conditions of soil moisture and plant nutrients make it difficult to establish vegetation, and the soil is most unsuitable for earth structures, while masonry or concrete structures are readily undermined or outflanked.

Piping or tunnel erosion

The formation of continuous pipes or channels underground is most common in those soil types subject to pinnacle erosion, but not entirely restricted to such soils. It occurs when surface water infiltrates through the soil surface and moves downwards until it comes to a less permeable layer. If there is a pressure gradient so that the water can flow laterally through the more porous soil and over the less permeable layer then the fine particles of the more porous soil may be washed out. The lateral flow then increases, and the pipe is enlarged, sometimes leading to roof collapse, so that the tunnel becomes a gully. An excellent review of the process is given by JONES (1981). Piping is less common on arable land than non-arable, except in parts of Australia, where deep ripping is used to break up the tunnels (BOUCHER 1990).

Mass movement and slumping

Slumping is usually a process of geological erosion, and although it may be accelerated, as in the sides of gullies, it can occur without any intervention of man. It can become the main agent in the development of gullies, as shown in cases where the head of the gully has worked right back up to the crest and beyond, where there can be no inflow at the head of the gully. These gullies probably often start due to flood flow in channels, but once the gully has started, erosion continues by slumping alone. The other main cases of slumping are river-bank collapse, and coastal erosion. Landslides, slip faults, and other geomorphological forms of mass movement are associated with saturated soils on steep slopes and unstable geological conditions (BRAND 1984). These are also mainly geological erosion processes, but the frequency or severity can be increased by changes in land use which increase the water content in the soil profile, for example where deforestation reduces the transpiration.

2.2.3 Degradation

The FAO/UNEP project, *Water Assessment of Soil Degradation*, listed six kinds of degradation: water erosion, wind erosion, excess of salts, chemical degradation, physical degradation, and biological degradation, but since it is difficult to quantify chemical, physical, and biological degradation, the study has so far been limited to the first three (FAO 1977). There are also some forms of degradation which do not involve any physical removal of soil.

Fertility erosion is the loss of plant nutrients by erosion, and can be comparable in magnitude with the removal of the same elements in the harvested crop (LAL 1987). The manner of the loss varies for different elements. Phosphorous is mainly lost when colloidal particles are eroded, but nitrogen can be lost in solution without any soil movement occurring.

The study of fertility loss has in recent years received much attention following the belated recognition that the significance of soil erosion is not the *quantity* of soil loss, but the *effect* of this loss on the productive capacity of the soil, especially through the loss of plant nutrients (STOCKING 1984, STOCKING and PEAKE 1987). There have been a number of workshops specifically on this topic (see *Further reading*) and there has been a flood of research studies on productivity, for example at recent ISCO conferences (listed in *Further reading*). An international network to remedy the lack of data on loss of soil nutrients has been set up by the FAO (STOCKING and SANDERS 1992). First attempts have been made to quantify the loss of productivity (RIJSBERMAN and WOLMAN 1984) and to model the process (BIOT 1988), and there will assuredly be further development on this topic.

Puddle erosion A comparable physical deterioration without net loss has been called puddle erosion because it can take place within a puddle. This is the physical breakdown of structure by rain, and the washing into depressions of the finer soil fractions, resulting in a structureless soil with a choked soil surface whose productive ability is much lowered (plate 2.5).

Vertical erosion Another physical translocation is the washing down of fine clay particles through a porous sand or gravel to accumulate at some less pervious layer lower down the profile. There are two effects possible; the loss of fine particles at one point and their increase at another point. In coarse sand soils an appreciable reduction in colloids and clays can result from vertical erosion with a consequent reduction of fertility. The effect where the fine material accumulates may also be undesirable when the result is the formation of a layer less permeable to roots and water.

Salinization Salinity is the condition when the soil contains an unduly high proportion of salts and it occurs for the most part in regions of arid or semi-arid climate as a result of geophysical redistribution and concentration of soluble salts. Secondary salinity results when the salts in saline soils are further redistributed by increased leaching and runoff resulting from farming operations. This is a serious form of land degradation in the drier parts of Australia, where it is known as *scald* or *seepage salinity* (plate 2.6) (BULLOCK and NEIL 1990).

2.2.4 *The relative importance of types of erosion*

The question is often debated as to which form of erosion is the most serious, and this is important if a conservation programme has

Plate 2.6 Ponding banks in order to leech salinity and encourage re-vegetation on 'scald' (Soil Conservation Service of New South Wales)

limited resources which are insufficient to tackle the whole erosion problem and must therefore be used against a selected part. However there is no single answer, for it depends upon the reasons why the erosion needs to be controlled. If the problem is that the production of food crops is jeopardized by erosion, then splash erosion and rill erosion on arable lands are the most important. If, however, the problem is off-site damage caused by a high sediment load interfering with storage dams for irrigation or hydroelectric schemes, or polluting water supplies, then the most important source of this silt will probably be gully erosion or streambank erosion. This is because the soil eroded by these forms goes immediately and wholly into the stream, whereas it is possible for soil to be lost in large quantities from arable lands and trapped in vegetation or deposited in ditches before it reaches the stream. The proportion of eroded material which reaches the stream is called *the delivery ratio.*

Plate 2.5 Puddle erosion is degradation of the soil by loss of structure without the soil being washed away (Copyright Aero Films Limited)

The importance of the different kinds of erosion, and the priorities given to their control, require an analysis of what the problem is, and what the objectives are of the remedial programme.

2.2.5 *Phases in the erosion process*

A similar situation arises when we consider the relative resistance to erosion of different soil types. Here different answers will be correct depending on what form of erosion is being considered. The soil might be resistant to surface erosion, but have a weak subsoil which makes it vulnerable to undercutting of streambanks. This is probably best illustrated by considering the classic divisions of the erosion process as propounded by Ellison nearly 50 years ago (ELLISON 1947). The three basic phases are detachment, transportation, and deposition, and Ellison showed in laboratory experiments that different soils rate differently in each of these phases. For example the particles of a fine sand are more easily *detached* than those of clay soil, but the clay particles are more easily *transported* than the sand particles. It is therefore necessary to define the nature of the erosion process when referring to how easily or how much the soil is subject to erosion.

Students of the processes of erosion refer to:

- *transport-limited erosion*, when there is not enough runoff to carry away the soil detached by raindrop splash. An extreme example would be a dry coarse sandy soil where there would be high infiltration and little runoff to move the detached particles
- *detachment-limited erosion* is when there is enough runoff to transport more soil than is actually being detached by splash. An example of this case might be gentle sustained rainfall on a saturated clay soil.

Readers who find it confusing to label the two situations by what *limits* the amount of erosion, might prefer the alternative which is to replace 'transport-limited' with 'detachment dominant', and 'detachment limited' with 'transport dominated'.

2.3 Calculating the amount of erosion

2.3.1 *Numerical estimations*

The essential qualifications of a science are the ability to define and measure the causes and effects of the natural phenomena related to the subject, and the ability to predict what will happen in given circumstances. Measuring the effects of erosion is fairly straightforward and measurements of the weight of soil lost have been carried out since the work of Wollny in the last century. As we saw in section 2.2.3, the effects have since been widened to include other aspects of degradation, some of which are more difficult to quantify. Earlier attempts at predicting erosion were hindered by a lack of understanding of the causes, and so were empirical, that is derived from observation and measurement of what happens, without necessarily understanding why it happens. The first 'black-box' model was *The Quantitative Evaluation of Factors in Water Erosion – A First Approximation* (MUSGRAVE 1947). During the next 30 years more data were accumulated from field plots of the experiment stations, and the equipment to store and retrieve the data steadily improved, as did the tools of statistical analysis (discussed in section 7.2.1). In the sixties there emerged efficient prediction models culminating in the Universal Soil Loss Equation (USLE) first published as a USDA handbook in 1965, and updated and revised in 1978 (WISCHMEIER and SMITH 1978). At the same time our understanding of the processes of erosion grew, and models based on the physical processes started to emerge. The causative agents for water erosion are rain and runoff, and theoretically each can contribute to the three erosion phases of detachment, transportation, and deposition. But the full model of three processes by two agents can be simplified.

Rain is high on detachment, low on transportation (except some downhill splash), and barely affects deposition. Runoff is low on detachment (except some scour if the velocity is high), and high on transportation and deposition. The most significant processes are therefore detachment by rainfall and transportation by runoff, and this is the basis for a number of process operated models such as EUROSEM and WEPP, as discussed in section 7.3. These models bring together in mathematical terms all the variable factors which influence erosion caused by rainfall. In subsequent chapters the factors will be separately considered in detail, but first we should establish a simple qualitative statement which puts into perspective the problems of erosion and erosion control.

2.3.2 *Qualitative statement of principles*

The fundamental cause of soil erosion is that *rain* acts upon the *soil*, and the study of erosion can be divided into how it will be affected by different kinds of rain, and how it will vary for different conditions of soil. The amount of erosion is therefore going to depend upon a combination of the power of the rain to cause erosion and the ability of the soil to withstand the rain. In mathematical terms Erosion is a function of the Erosivity (of the rain) and the Erodibility (of the soil), or *Erosion=f (Erosivity) (Erodibility)*.

Erosivity can be defined as the potential ability of the rain to cause erosion, and for given soil conditions one storm can be compared quantitatively with another and a numerical scale of values of erosivity can be created. *Erodibility* is defined as the vulnerability of the soil to erosion, and for given rainfall conditions, one soil condition can be compared quantitatively with another, and a numerical scale of values of erodibility can be created. *Erodibility of the soil* can be subdivided into three parts. Firstly the fundamental or inherent characteristics of the soil – its mechanical, chemical, and physical composition – the things which can be measured in the laboratory, as discussed in chapter 5. Secondly the topographic features, especially the slope of the land. Thirdly the erodibility will also depend on what treatment is given to the soil, i.e. how it is managed. This management may in turn be subdivided into two parts, land management, and crop management. The broad issues, which can be grouped under the heading land management, are the kind of land use, ie forestry, grazing, arable, etc. These are discussed in chapter 9. The subsequent decisions are more detailed. For arable land they include the kind of crop, the fertilizer treatment, the harvesting, and so on. These all comprise *crop management* and are discussed in chapter 11. Some conservation management practices, such as contour ploughing or terracing, are bound up with both the broader issues of land management and the mechanics of crop management. Figure 2.2 shows this approach diagrammatically, and leads to combining all the factors which influence erosion into the Universal Soil Loss Equation. In subsequent chapters the factors will each be analyzed in detail.

References

BIOT, Y. J. 1988 Modelling Productivity Losses Caused by Erosion, in *Land Conservation for Future Generations*, edited by Sanarn Rimwanich, Department of Land Development, Bangkok

BOUCHER, S. C. 1990 *Field Tunnel Erosion: Its Characteristics and Amelioration*, Department of Geography, Monash University, Victoria, Australia

BRAND, E. W. 1984 *Landslides in Southwest Asia: A State-of-the-Art Report*, Proceedings of 4th International Symposium on Landslides, Toronto, Vol 1, 17–59

BULLOCK, P. R. and D. T. NEIL 1990 The Catastrophic Nature of Seepage Scald Formation in Southeastern Australia, in *Catena*

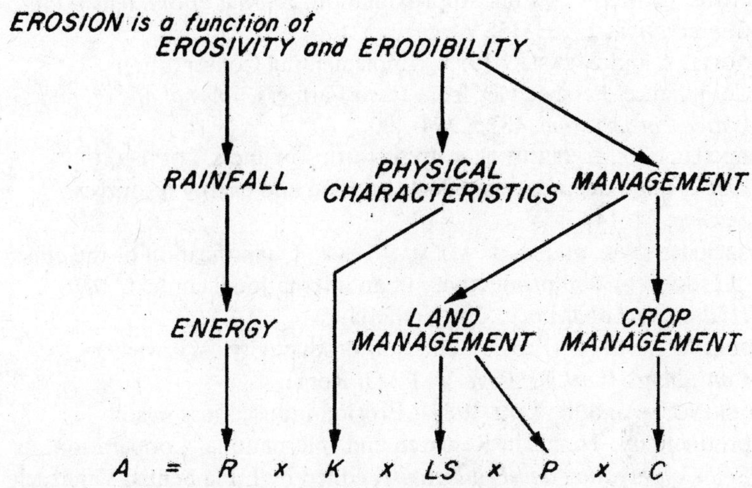

Figure 2.2 The factors which affect rainfall erosion. The Universal Soil Loss Equation $A = R \times K \times LS \times P \times C$ is discussed in chapter 7

Supplement 17, edited by R. B. Bryan, Catena Verlag, Cremlingen, Germany

EDWARDS, M. 1961 Soil Erodibility Factors and Soil Loss Tolerance, in *Soil Loss Prediction for Arkansas, Louisiana, Mississippi, Oklahoma and Texas*, USDA-SCS, Fort Worth, Texas

ELLISON, W. D. 1947 Soil Erosion Studies, Parts I to VII, *Agricultural Engineering*, 28, 145, 197, 245, 297, 349, 402, 442

FAO 1977 Assessing soil degradation, *Soils Bulletin 34*, FAO, Rome

JOHNSON, L. C. 1987 Soil Loss Tolerance: Fact or Myth? *Journal of Soil and Water Conservation*, 42, 3, 155–160

JONES, J. A. A. 1981 *The Nature of Soil Piping – a review of research*, BGRC Research Monograph, Geo Books, Norwich, Norfolk

LAL, R. 1987 Effects of soil erosion on crop productivity, *Critical Reviews in Plant Science*, 5, 4, 303–367

MANNERING, J. V. 1981 The Use of Soil Loss Tolerances as a Strategy for Soil Conservation, in *Soil Conservation: Problems and Prospects*, edited by R. P. C. Morgan, Wiley, Chichester, Sussex

MORGAN, R. P. C. 1986 *Soil Erosion and Conservation*, Longman Scientific and Technical, Harlow, Essex

MUSGRAVE, G. W. 1947 The Quantitative Evaluation of Factors in

Water Erosion – A First Approximation, *Journal of Soil and Water Conservation*, 2, 133–138

PADGITT, S. and P. LASLEY 1993 Implementing Conservation Compliance: Perspectives from Iowa Farmers, *Journal of Soil and Water Conservation*, 48, 5, 394–400

PIMENTEL, D., E. C. TERHUNE, R. DYSON-HUDSON and S. ROCHERAU 1976 Land degradation: effects on food and energy resources, *Science*, 94, 149–155

RIJSBERMAN, F. R. and M. G. WOLMAN 1984 Quantification of the effect of erosion on soil productivity in an international context, *Delft Hydraulics Laboratory*, The Netherlands

STOCKING, M. 1984 Erosion and Soil Productivity: A Review, *Consultants Working Paper 1*, FAO, Rome

STOCKING, M. and L. PEAKE 1987 Erosion-induced loss in soil productivity: Trends in Research and International Cooperation, in *Soil Conservation and Productivity*, edited by I. Pla Sentis, Venezuela Soil Science Society, Maracay

STOCKING, M. and D. W. SANDERS 1992 The Impact of Erosion and Soil Productivity in *People Protecting their Land*, edited by P. G. Haskins and B. M. Murphy, Department of Conservation and Land Management, Sydney, New South Wales

WISCHMEIER, W. H. and D. D. SMITH 1978 Predicting Rainfall Erosion Losses – A Guide to Conservation Planning, *Agricultural Handbook 537*, USDA, Washington, DC

3

The physics of rainfall

3.1 Physical characteristics

There is obviously an association between the amount of rainfall and the amount of soil erosion, i.e. more rain goes with more erosion, and less rain with less erosion, but in statistical terms the correlation between the two is poor. The same total quantity of rain can on different occasions result in widely differing amounts of erosion, and so other more specific measures are required to describe the ability of rainfall to cause erosion. In this chapter we shall establish the known facts about the main physical properties of rainfall, and then in chapter 4 show how these properties are related to the erosive power of rainfall.

3.1.1 Quantity of rainfall

Any measure of rainfall amount is a sample, and so associated with the inevitable problems of sampling, namely, 'is the sample representative of the whole?' and 'is the sample measured accurately?' On both counts the data of rainfall which are usually available are of very doubtful reliability. Considering firstly the size of the sample measured, a very intensive network of rain gauges such as that established in an experimental catchment might be as dense as one gauge per 25 hectares, but this corresponds (with a 125 mm diameter gauge) to a sample of about one in 20 million, which would for any other scientific measurement be considered totally inadequate. The density of gauges in a country which considers itself very well provided might be one gauge per 25 square kilometres – a sample of one in 2 000 million, and in an undeveloped country the density of one gauge per 2 500 square kilometres is a sample of 1 in 200 000 000 000! Secondly, the siting of this remarkably small number of gauges is usually arranged primarily on convenience, *eg* where they can be easily read and contained, with little or no consideration of whether the site is representative of the whole. Thirdly, the accuracy of the measurement is usually unknown and studies have shown that variations in the shape and size of the instrument, and in the height at which it is mounted, and the way it is shielded from the wind, can seriously affect the recorded amount of rainfall. Clearly, even the apparently straightforward question of how much rain falls is not answered as precisely as might be imagined from a survey of rainfall tables and maps.

In addition to the straightforward measurement of the quantity of rain, there are a number of other questions to consider. If a rainfall map shows an average annual rainfall of say 700 mm this implies that if a given area, say a square metre, accumulated the year's rain it would rise to a depth of 700 mm over this area. If the ground is sloping, then what looks like a square metre on the map will be a considerably larger area of actual ground surface. The volume of water has to spread over the larger area and so the effective rainfall is considerably reduced. The problems of measuring rainfall on slopes and in remote areas are thoroughly presented by HAMILTON (1954).

Another aspect of the effectiveness of rain is how it is distributed through the year. The fairly uniform spread which is common in temperate climates is favourable for intensive agriculture and least conducive to erosion. In tropical climates it is more usual for the rain to be concentrated into part of the year. Erosion is more likely to result from both the heavier rain in the wet season and the desiccation of plant life during the dry season. A single rainy season is described as a *uni-modal distribution* (figure 3.1). Another distribution pattern common in the sub-tropics is the *bi-modal*, i.e. the rainy season is divided into two parts with a dry period between. These are often described as the little rains and the big rains, and if the dry spell in between allows harvesting or tillage operations this distribution can be agriculturally desirable.

3.1.2 *Intensity or rate of rainfall*

There is considerable evidence of a close association between erosion and intensity, and intensity is particularly important as a potential parameter of erosivity because it is the only feature of rainfall which, in addition to amount, is frequently recorded at conventional meteorological stations. Data on occurrence and frequency of intensities are nowadays recorded on most meteorological stations, and no erosion experiments would be properly equipped without some form of intensity recorder (plate 3.1). Any erosivity index based solely, or primarily, on some function of intensity has a greater scope for its application than an

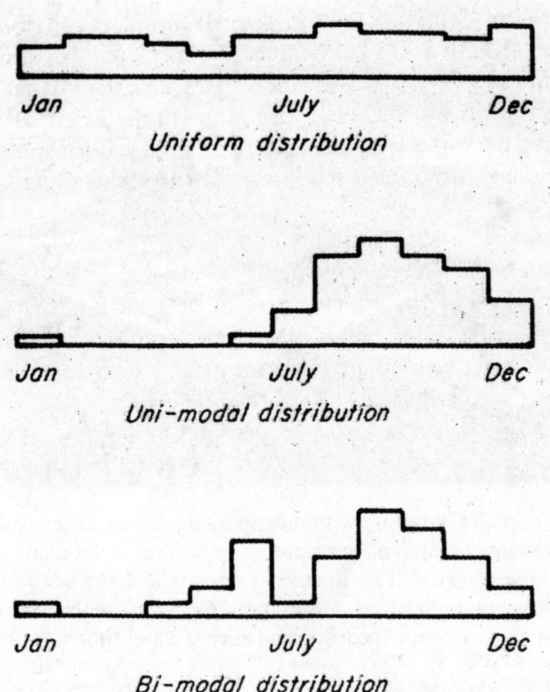

Figure 3.1 Patterns of distribution of annual rainfall

index based on any other rainfall characteristic. An index based on, for example, kinetic energy would be extremely useful even if it required new instrumentation to measure the kinetic energy, but would only be applicable after the collection of new data, and so less useful than an index based on intensity.

Measurements of intensity sufficient for meteorological purposes are normally obtained from recording rain gauges where successive increments of rainfall are recorded as a cumulative total on a clock-driven chart whose speed varies according to the frequency with which the chart is to be changed, usually daily, or in remote or inaccessible areas, weekly or monthly. The intensity is computed from the rate of change in the quantity of rainfall recorded, i.e. the gradient of the recorded line. This indirect measurement is suitable for averaging the intensity over fairly long periods, but when intensity data are required for short time intervals the method becomes very laborious, and it is not very accurate (figure 3.2).

Instruments which directly record the rate of rainfall rather than the

Plate 3.1 A rainfall laboratory with measuring devices arranged round an underground room containing the recording apparatus. *A* Standard rain gauge, *B* Directional rain gauge, *C* Two intensity recorders, *D* Drop size sampling machine, *E* Momentum balances, *F* Soil trays. Not seen are the recording rain gauges, splash cups, acoustic recorder, and energy wheel (HUDSON 1965)

quantity are available, and are often used in connection with soil erosion experiments, but are seldom included in the equipment of ordinary meteorological stations.

In temperate climates the rainfall rate seldom exceeds 75 mm per hour, and then only in summer thunderstorms. In many tropical countries intensities of 150 mm per hour are experienced regularly. A maximum rate, sustained for only a few minutes, was recorded by the author in Africa at 340 mm per hour.

3.1.3 Raindrop size

The earliest recorded measurements of the size of raindrops were made in 1892 by LOWE who caught raindrops on flat sheets of slate which were ruled off into squares so that the size of the splashes could be measured. This method of calculating the size of a drop from the size of the splash which it makes on some collecting device is one of the most popular methods. One technique is to take absorbent paper and lightly dust onto the surface a very finely powdered water-soluble dye. In the dry state the dye is visible but on exposure to rain each raindrop make a roughly circular stain which can be measured later. The drop size can be

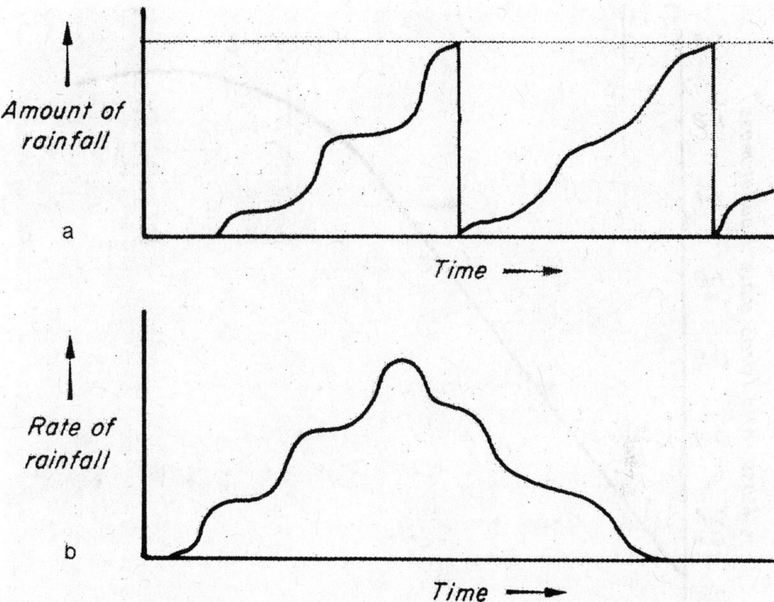

Figure 3.2 Measurements of intensity (*a*) Indirect. The slope of the line on an autographic rain gauge represents the rate of rainfall. (*b*) Direct. Intensity is measured directly, and the amount of rain is the area under the line

calculated from the formula $D = aS^b$, where D is the drop diameter, S is the stain diameter, and a and b are constants established by laboratory calibration for the paper used. The method is reviewed by HALL (1970) and is increasingly attractive now that the previously tedious job of measuring and counting the drop stains can be accomplished by the use of an electronic scanner and image analysers (ATTLE *et al* 1980).

Another popular approach is the *Flour-Pellet* method (HUDSON 1964a). A sample of rain is caught in a dish containing flour. Each raindrop forms a small globule of wet flour and when dried in an oven these set into hard pellets which can be separated from the rest of the flour. A laboratory calibration has previously established the relationship between the size of a pellet and the size of the drop which formed it (figure 3.3).

The size and size distribution of raindrops may also be measured by some of the pressure transducers used to measure rainfall impact as described in section 3.2.

Using such methods (and some of the others described in HUDSON (1964b) the range of raindrop sizes has been measured in many different

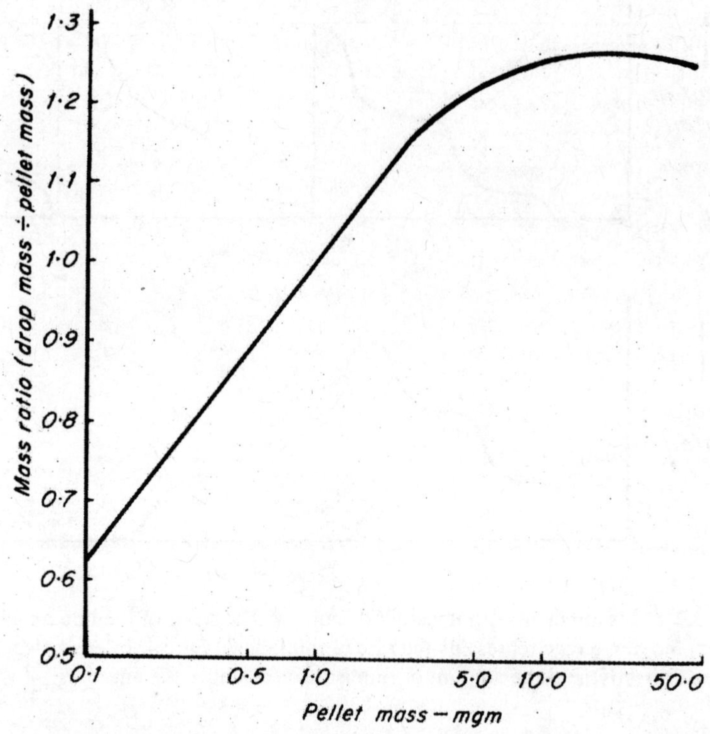

Figure 3.3 Calibration for the flour-pellet method of measuring the size of raindrops

countries and in many different types of rainfall. The upper limit appears to be about 5 mm diameter and drops bigger than this break up into a number of smaller drops. BLANCHARD (1950) showed in wind tunnel experiments that drops are stable up to 4.6 mm diameter, and unstable above 5.4 mm diameter with a range in between where the drops may or may not disintegrate depending on the turbulence. High speed photographs show that the shape of falling drops is not at all like the conventional tear drop shape, but a sphere well flattened by the air resistance. Field observations agree with these maximum sizes. Drops are only infrequently found larger than 5 or 6 mm diameter, but do occasionally occur. This probably happens when two drops have collided to form a 'super drop' which reaches the ground before it has time to split up again.

3.1.4 *Drop size distribution*

Since the rain is made up of all sizes, it is also necessary to determine the proportions of large and small drops (i.e. the size distribution) and how this distribution varies in different kinds of rain. It is clear from everyday observation that the low intensity rain which can last for days is mainly made up of small drops, whereas the high intensity rain from a thunderstorm has at least some drops which are much bigger. In both the *drop-stain* and the *flour-pellet* method of measuring drop size a sample of the rain is collected and so the distribution of drop sizes may be determined. The classical study in this field of work was that carried out by LAWS and PARSONS (1943). One of the difficulties of sampling rainfall is that of catching those storms which only occur infrequently, and for many years our knowledge of drop size distribution was limited to low intensity rain. Later studies at high intensities significantly change the picture.

It is not easy to describe a distribution by a single parameter. The 'average' drop size gives little indication of how the average is made up, and probably the best index for drop distributions is the median volume drop diameter (D_{50}). It is obtained from a plot of cumulative volume against drop diameter (figure 3.6) when half of the volume of rain falls in drops with a smaller diameter, and half as bigger drops. Studies such as that of BEST (1950) showed that the relationship between D_{50} and intensity I is of the form $D_{50} = aI^b$ (a and b are constants), and there is no doubt that this form, describing increasing size with increasing intensity, is valid for low intensities. However, we know that there is a physical upper limit to the maximum drop size, and studies of high intensity rain (HUDSON 1963) showed that there is in fact a reversal of this relationship at very high intensities. Figures 3.4 and 3.5 show drop size distribution curves for increasing intensity and it is clear that the modal value of drop diameter (i.e. the peak of the curve) rises until about 80 or 100 mm per hour but then decreases at still higher intensities. Plotting D_{50} against intensity (figure 3.7) shows that the form $D_{50} \propto I^b$ cannot be extrapolated to high intensity tropical rainfall. It is not likely that there is a single unique relationship between D_{50} and intensity; more probably it varies for different kinds of rain, but the principle, first identified in Zimbabwe, that D_{50} levels off and possibly falls slightly is now reinforced by evidence from Thailand (BARUAH 1973), from Nigeria (KOWAL and KASSAM 1977), and from the USA (CARTER *et al* 1974). Even in tropical rainfall the intensity does not often exceed 150 mm per hour, but it is important to understand the nature of high intensity rain, and how it differs from temperate rain.

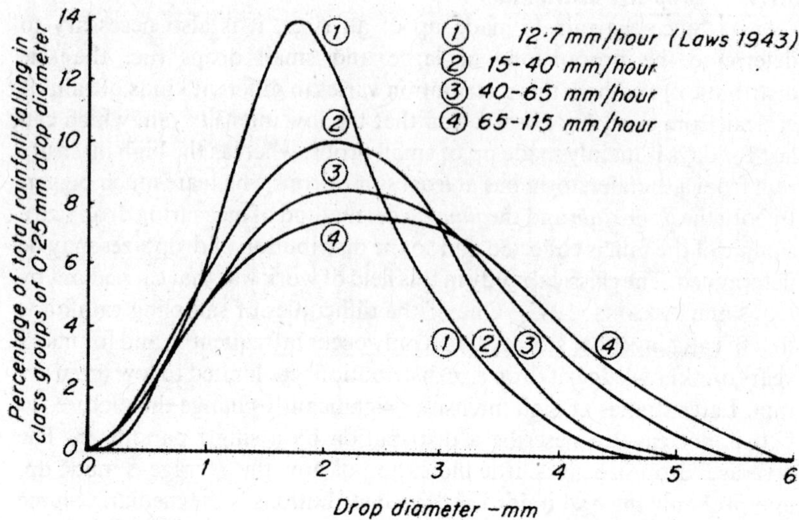

Figure 3.4 Drop size distribution at low and medium intensities

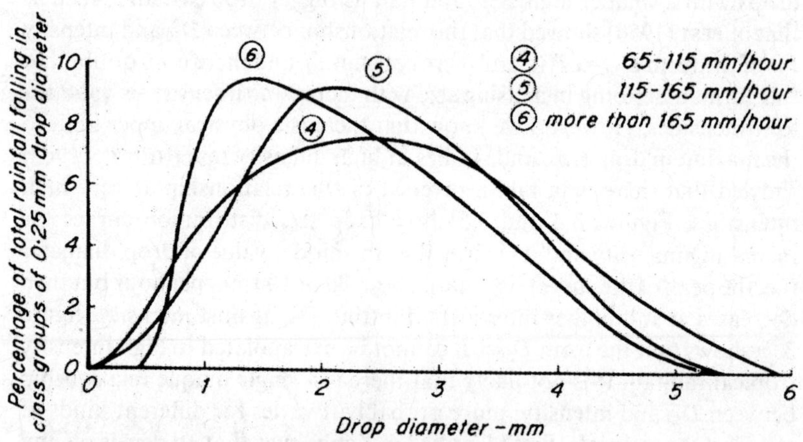

Figure 3.5 Drop size distribution at high intensities

3.1.5 *Terminal velocity*

A body falling freely under the force of gravity will accelerate until the frictional resistance of the air is equal to the gravitational force, and will then continue to fall at that speed. This is known as the *terminal velocity* and depends upon the size and shape of a body. The terminal

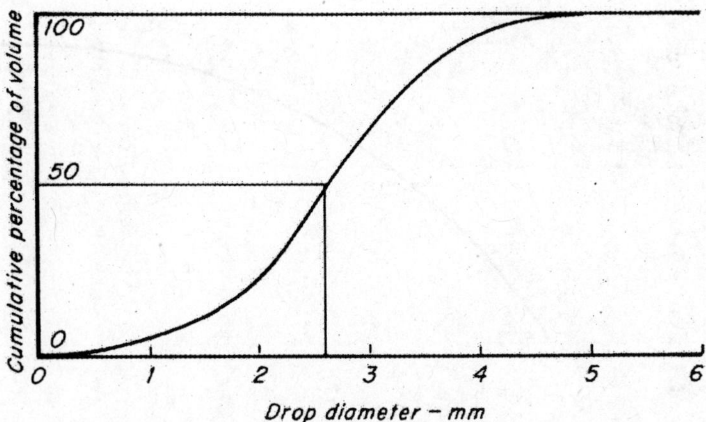

Figure 3.6 Finding the median volume drop diameter (D_{50})

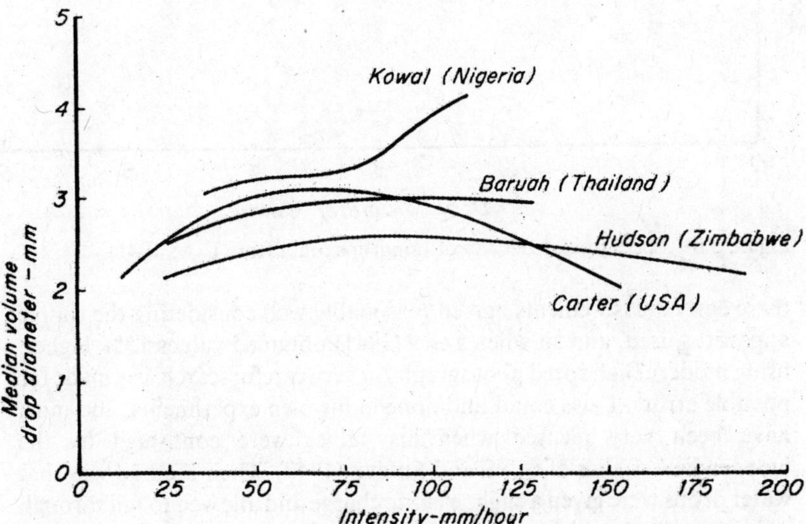

Figure 3.7 The relation between median volume drop diameter and intensity

velocity of raindrops increases as the size increases, the largest drops of about 5 mm diameter having a terminal velocity of about 9 metres per second (figure 3.8).

Many laboratory measurements of falling drops of water were carried out by physicists at the beginning of this century, particularly by Mache, Schmidt, Liznar, and Lenard in Germany, and Flower in England. All

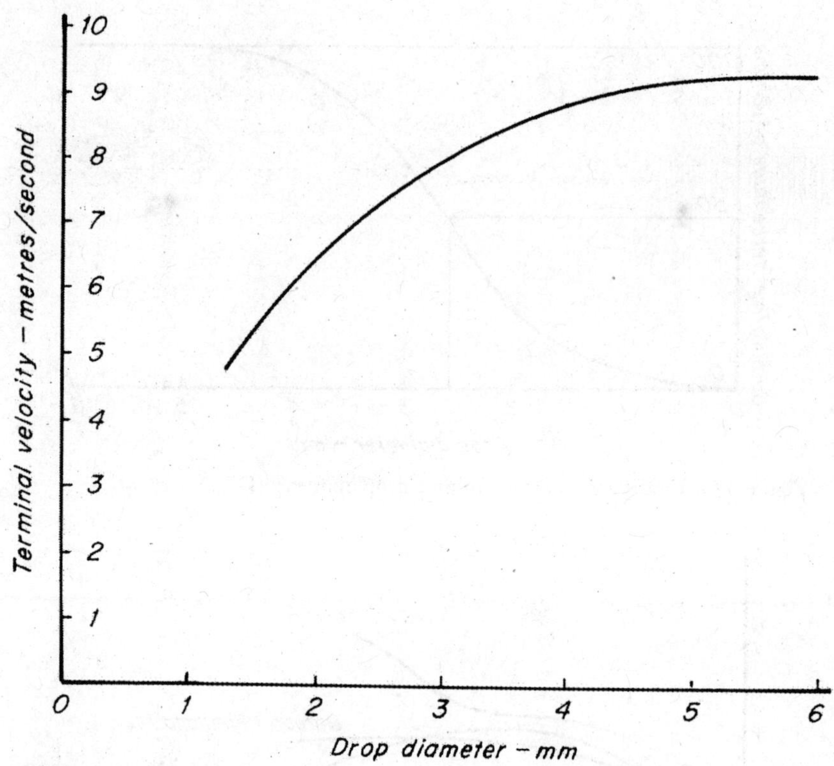

Figure 3.8 The terminal velocity of raindrops (data from LAWS 1941)

these early measurements agreed reasonably well considering the simple apparatus used, and so when LAWS (1941) obtained values 15% higher, using modern high speed photography, a very careful search was made for possible errors. Laws could find none in his own experiments, and must have been very pleased when his values were confirmed by the independent studies of GUNN and KINZER (1949). Their method was that water drops were given a slight electric charge and allowed to fall through induction rings which gave an electrical impulse when the drop fell through them. The impulses were fed through amplifiers to an oscillograph so that the time to fall the measured distance between the two induction rings could be measured very accurately. Gunn and Kinzer's results differed from those of Laws by less than 3%, and have more recently been confirmed by HINKLE *et al* (1987). Laws also measured the terminal velocity of raindrops in the open, and found that while turbulence and wind have some effect, nevertheless most raindrops reach the ground at 95% of their 'still-air' terminal velocity. When rain is accompanied by wind there is an added sideways component of velocity, and the resultant vector may be greater than the still-air velocity. The effect will be greater on small drops falling slowly than on large drops with higher velocity. There is some evidence that in tropical rainfall highest rates are likely to occur in a relatively still air (HUDSON 1964c) and this may be connected with the wind patterns around the base of convective thunderstorms.

3.2 Momentum and kinetic energy

There is experimental evidence that the erosive power of rainfall is related to compound parameters derived from combinations of more than one physical property. The kinetic energy of the rain and its momentum are examples. If the size of raindrops is known and also their terminal velocity, it is possible to calculate the momentum of the falling rain, or its kinetic energy, by a summation of the values for individual raindrops. In the past this indirect calculation method has given better results than attempts to measure directly the momentum of kinetic energy of falling rain. The forces involved are so small that any instrument sufficiently sensitive to record them mechanically is liable to be swamped by wind effects. However, developments in technology have introduced new possibilities for converting the energy of rainfall into other forms which may be more easily measured.

Acoustic sensors In this method the noise of the rain falling on a diaphragm is picked up by a microphone which gives a measurable signal. The advantage is that the sound signal is integrated across the whole of the raindrop spectrum (KINNELL 1972). Also by tuning the circuit this method can be adjusted to measure intensity, or momentum, or kinetic energy (HUDSON 1965). The sensor can be quite large so that a representative sample of rainfall can be measured.

The disadvantage of the acoustic method is that a drop striking the centre of the diaphragm may produce a different effect compared with a drop striking near the edge of the diaphragm. This can be partially overcome by shielding all but the centre portion. Another disadvantage is that it is difficult to manufacture uniform diaphragms which do not require individual calibration.

Pressure transducers Several instruments have been developed to measure momentum by recording the physical displacement of a target sensor against an elastic spring (NEAL and BAVER 1937), or against gravity (ROSE 1958, and HUDSON 1965). The same principle is more elegantly used nowadays by the use of bonded strain gauges (WEBSTER 1980).

Alternatively the target sensor could be a diaphragm pressure transducer and it would seem feasible to use the unbonded strain gauge type, or the linear variable displacement type, or capacitance, or semi-conductors (HENRY 1975). These could all be classed as high-inertia pressure transducers and are fairly sophisticated, and at the present time expensive. Their use will probably be limited to research instruments rather than for widespread use within a meteorologic network.

Piezoelectric sensors These devices are used in record players, where changes of pressure on quartz crystal generate an electric signal. The principle of the alternative kind of record player cartridge, the magnetic head, is used in the commercially available Joss-Waldvogel Distrometer (JOSS and WALDVOGEL 1967, KINNELL 1976).

A piezoelectric sensor used with unsophisticated sorting and storage equipment gave important results in Nigeria (KOWAL and KASSAM 1977). Another simple but effective application is that of HEUVELDOP and KRUSE (1978). Several other similar applications are currently being developed. Current developments in pressure transducers which could be used to sense the raindrop impact, and in microprocessors to grade, sort, and store the data suggest that a new generation of instruments will soon emerge (HUDSON 1981).

Earlier studies have been carried out in widely separated countries and some of the main results are shown in figure 3.9. One difficulty is that few measurements have been made at high intensities; another is that these studies were made by independent research workers, and there is no way of telling how much variation is due to their different techniques and how much is real differences between the rain in various countries. The results of some of these studies have been presented as mathematical equations, compared in table 3.1. All the investigators report very wide scatter, and it is questionable whether the data really justify the precision of some of the equations.

Research in this subject is being stimulated by the knowledge that there is a strong connection between the momentum or energy of rain and its power to cause erosion. The consideration of why some storms can cause more erosion damage than others, and ways in which this erosive power can be determined are the subject of chapter 4.

Table 3.1 Equations relating kinetic energy and intensity

	Units: *Energy*	Units: *Intensity*
1 $E = 916 + 331 \log I$ which converts to	foot-tons/acre-inch	in/h
2 $E = 210 + 89 \log I$ or to	tonne-metres/hectare-cm	cm/h
3 $E = 11.9 + 8.7 \log I$	J/m^2-mm	mm/h
This relationship has also been expressed in $kg\text{-}m/m^2$ and in megajoules/ha		
4 $E = 29.22(1 - 0.894e^{-0.0477I})$	J/m^2-mm	
5 $E = 0.29(1 - 0.596e^{-0.041I})$	MJ/ha.cm	cm/h
6 $E = 30 - \frac{125}{I}$	J/m^2-mm	mm/h
7 $E = 9.81 + 11.25 \log I$	J/m^2-mm	mm/h

SOURCES: 1 WISCHMEIER *et al* 1958, 2 WISCHMEIER and SMITH 1978, 4 KINNELL 1981, 5 ROSEWELL 1986, 6 HUDSON 1965, 7 ZANCHI and TORRI 1981

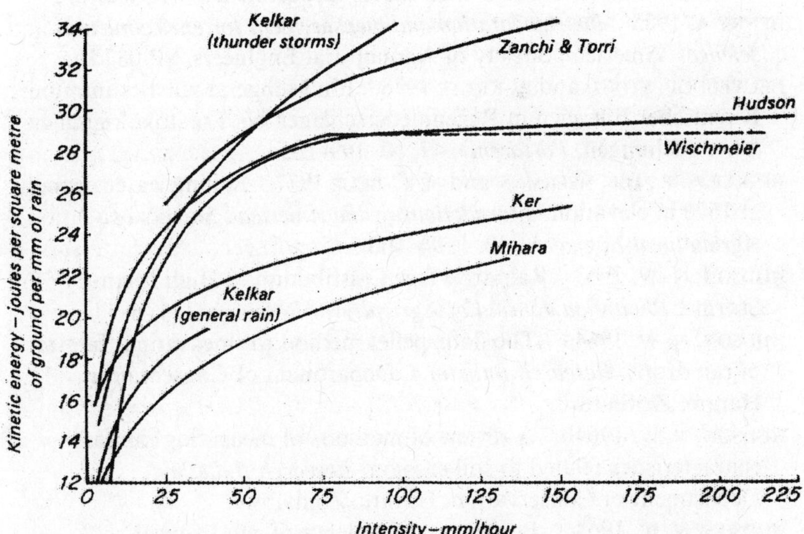

Figure 3.9 The relation between kinetic energy of rainfall and intensity. The full lines extend to the highest intensity recorded. The studies were carried out in different countries. HUDSON–Zimbabwe, KELKAR–India, KER–Trinidad, MIHARA–Japan, WISCHMEIER–United States, ZANCHI and TORRI–Italy

References

ATTLE, J. R., D. ONEY and R. A. SWENSON 1980 Applications of image analysis, *American Laboratory*, April 1980

BARUAH, P. C. 1973 *An Investigation of Dropsize Distribution of Rainfall in Thailand*, Master's Degree Thesis 528, Asian Institute of Technology, Bangkok

BEST, A. C. 1950 The Size Distribution of Raindrops, *Quarterly Journal of the Royal Meteorological Society*, 76, 16

BLANCHARD, D. C. 1950 Behaviour of Water Drops at Terminal Velocity, *Transactions of the American Geophysical Union*, 31, 836

CARTER, C. E., J. D. GREER, H. J. BRAUD and J. M. FLOYD 1974 Raindrop Characteristics in South Central United States, *Transactions of the Association of Agricultural Engineers*, 17, 6, 1033–1037

GUNN, R., and G. D. KINZER 1949 Terminal Velocity of Water Droplets in Stagnant Air, *Journal of Meteorology*, 6, 243

HALL, M. J. 1970 Use of the stain method in determining the dropsize distribution of coarse liquid sprays, *Transactions of the Association of Agricultural Engineers*, 13, 33–37, 41

HAMILTON, E. L. 1954 Rainfall Sampling on Rugged Terrain, *Technical Bulletin 1096*, United States Department of Agriculture

HENRY A. 1975 *Instrumentation and measurement for environmental sciences*, American Society of Agricultural Engineers, SP-0375

HEUVELDOP, VON J. and R. KRUSE 1978 Ein Mebgerat zur Bestimmung Kinetischer Energien in Regenniederschlagen bei Forstokologischen Untersuchungen, *Fortarchiv*, 49, 10, 199–203

HINKLE, S. E., D.F. HEERMAN and M. C. BLUE 1987 Falling water drops at 1570 m elevation, *Transactions of the American Society of Agricultural Engineers*, 30, 1, 94–100

HUDSON, N. W. 1963 Raindrop Size Distribution in High Intensity Storms, *Rhodesian Journal of Agricultural Research*, 1, 1, 6–11

HUDSON, N. W. 1964a The flour-pellet method for measuring the size of raindrops, *Research Bulletin 4*, Department of Conservation, Harare, Zimbabwe

HUDSON, N. W. 1964b A review of methods of measuring rainfall characteristics related to soil erosion, *Research Bulletin 4*, Department of Conservation, Harare, Zimbabwe

HUDSON, N. W. 1964c Bearing and incidence of sub-tropical convective rainfall, *Quarterly Journal of the Royal Meteorological Society*, 90, 385, 325–328

HUDSON, N. W. 1965 *The influence of rainfall on the mechanics of soil erosion*, MSc Thesis, University of Cape Town

HUDSON, N. W. 1981 Instrumentation for studies of the erosive power of rainfall, in *Proceedings of Symposium on Erosion and Sediment Transport Measurement*, Florence, Italy, International Association of Hydrological Sciences publication 133, 383–390

JOSS, V. J. and A. WALDVOGEL 1967 Ein spectrograph für Niederschlagstropher mit automatischer Auswertung, *Pure and Applied Geophysics*, 68, 240–246

KINNEL, P. I. A. 1972 The acoustic measurement of water drop impacts, *Journal of Applied Meteorology*, 11, 691–694

KINNEL, P. I. A. 1976 Some observations of the Joss-Waldvogel rainfall disdrometer, *Journal of Applied Meteorology*, 15, 499–502

KOWAL, J. M. and A. H. KASSAM 1977 Energy load and instantaneous intensity of rainstorms at Samaru, Northern Nigeria, in *Soil Conservation and Management in the Humid Tropics*, edited by D. J. Greenland and R. Lal, Wiley, Chichester, Sussex

LAWS, J. O. 1941 Measurements of fall-velocity of water-drops and raindrops, *Transactions of the American Geophysical Union*, 24, 452

LAWS, J. O. and D. A. PARSONS 1943 The relation of raindrop size to intensity, *Transactions of the American Geophysical Union*, 24, 452–460

NEAL, J. H. and L. D. BAVER 1937 Measuring the impact of raindrops, *Journal of American Society of Agronomics*, 29, 708–709

ROSE, C. W. 1958 *Effects of rainfall and soil factors on soil detachment*, PhD Thesis, London University

WEBSTER, D. K. 1980 *An investigation into the use of microprocessors for the measurement of rainfall intensity*, BSc Dissertation H/80/197, National College of Agricultural Engineering

4

The erosivity of rainfall

4.1 Defining erosivity and erodibility

Rainfall erosion is the interaction of two items – the rain and the soil. The amount of erosion which occurs in any given circumstances will be influenced by both, and our study of the processes of soil erosion is simplified by considering the two aspects separately.

We know from observation that one storm can cause more erosion than another on the same land, and we also know that the same storm will cause more erosion on one field than on another. The effect of the rain is called *erosivity* and the effect of the soil is called *erodibility*. The way they differ is best explained by examples.

Let us imagine an experiment station which has over its whole area exactly the same soil type and slope. Assume also that the whole of the station is covered by a completely uniform crop. If the amount of erosion taking place during each storm were measured, the amount would be influenced only by the nature of the rain. Comparing the effect of one storm with that of another storm would provide a relative measure of the power of each storm to cause erosion, and that is what is meant by erosivity. This will enable us to say that on this imaginary experiment station, storm *A* caused a soil loss of *X* kilograms per hectare and storm *B* caused a soil loss of *Y* kilograms per hectare. If *Y* is half of *X*, then the erosivity of storm *B* is half of the erosivity of storm *A*.

But also we find that on a farm a few miles away the soil is different, and during storm *A* the soil loss is 2*X* kilograms per hectare. The difference is because one soil is more easily eroded than the other, and this vulnerability to erosion is what is meant by erodibility. It is thus the reciprocal of the soil's resistance to erosion. This example is shown diagrammatically in figure 4.1.

It is usual to assume that if one soil is eroded twice as much as another

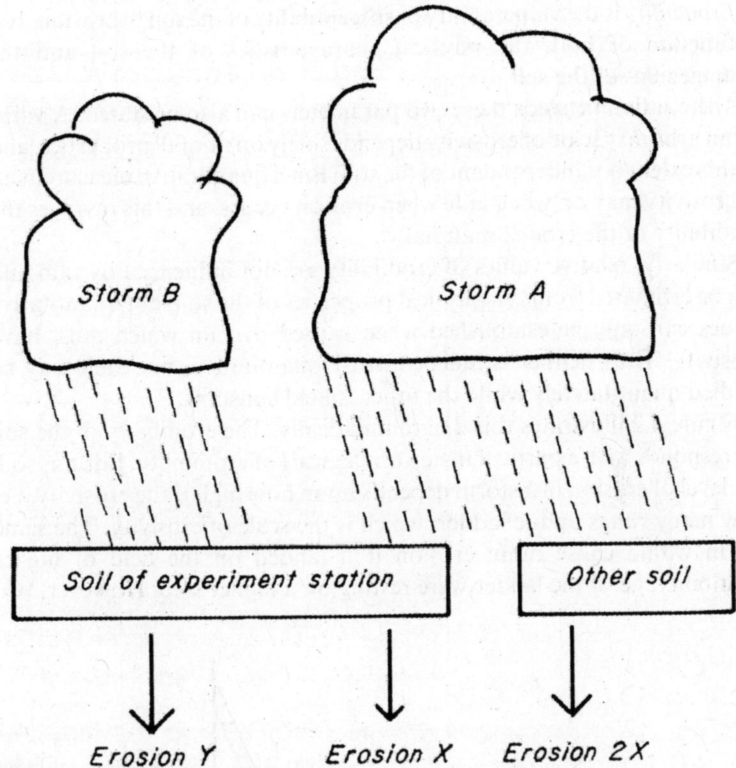

Figure 4.1 The erosivity of rainfall and the erodibility of soil

soil during one storm, it will lose twice as much in all storms, that is we are assuming that there is no interaction between erosivity and erodibility. This is not necessarily true, but the assumption is condoned because it greatly simplifies the quantitative study of the erosion process.

The resistance of a soil to erosion depends on many factors, so to measure erodibility numerically an assessment has to be made of each factor. Some of these are *the nature of soil* (sandy soil is more easily eroded than a hard clay), *the slope of the land* (a steep slope will erode more than a flat one) and *the kind of crop* (some crops protect the soil better than others). The evaluation of the factors affecting the erodibility of the soil will be considered in chapter 5.

Formal definitions can now be stated as follows:

Erosivity is the potential ability of rain to cause erosion. It is a function of the physical characteristics of rainfall.

Erodibility is the vulnerability or susceptibility of the soil to erosion. It is a function of both the physical characteristics of the soil and the management of the soil.

The relation between these two parameters can also be stated: A value on an arbitrary scale of erosivity depends solely on rainfall properties, and to this extent it is independent of the soil. But a quantitative measurement of erosivity may only be made when erosion occurs, and this involves the erodibility of the eroded material.

Similarly, relative values of erodibility are not influenced by rain and can be estimated from the physical properties of the soil, but quantitative values can only be established when caused by rain which must have erosivity. Thus neither is independently quantitative, but each may be studied quantitatively while the other is held constant.

Figure 4.2 illustrates this diagrammatically. The erodibility of the soil corresponds with a particular step on the scale of erodibility. For this soil, the level of erosion in a storm depends upon how high is the erosivity, i.e. how many rungs up the ladder, which is the scale of erosivity. The same storm would cause more erosion if it landed on the field of higher erodibility, i.e. if the ladder were resting on a higher step. However, two

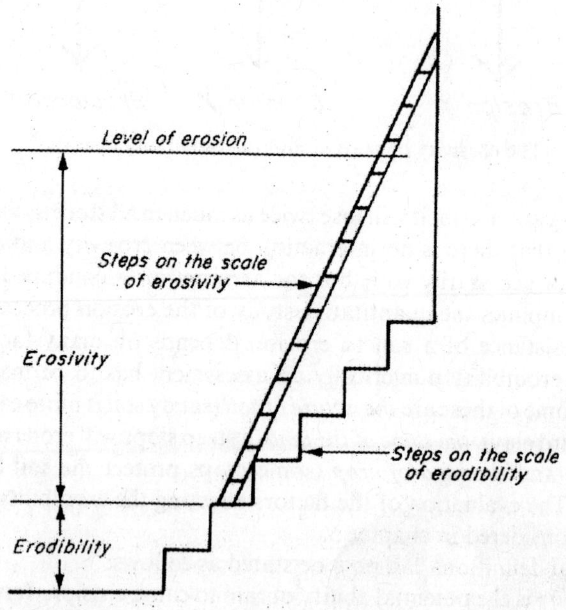

Figure 4.2 The combination of erosivity and erodibility

caveats must be applied to this analogy. Firstly, we are assuming that erosivity and erodibility are completely independent. Actually it is possible that they may interact, in which case we should use ladders with different spacing between the rungs to stand on the different steps of erodibility. Secondly, we shall see in chapter 7 that numerical values of erosivity and erodibility are commonly combined by multiplying instead of adding them together.

Definitions of erosivity and erodibility should specify the form of erosion. Rainfall characteristics which cause splash erosion will not be the same as those which cause gully erosion. In the rest of this chapter, and in chapter 5, erosivity and erodibility will be taken to relate to inter-rill and rill erosion, and we will continue to treat them as if they were independent.

4.2 Erosivity, energy, and intensity

4.2.1 Raindrop splash and surface runoff

Soil erosion is a work process in the physical sense that work is the expenditure of energy, and energy is used in all the phases of erosion – in breaking down soil aggregates, in splashing them in the air, in causing turbulence in surface runoff, in scouring and carrying away soil particles (plate 4.1). If the available sources of energy are considered we can see why splash erosion is so vital in the erosion process. In table 4.1 the kinetic energy available in falling rain is compared with that from surface runoff. The actual figures used in this calculation are not important since they are based on assumptions of the percentage runoff and assumed velocities, but clearly the difference in the amounts of energy is very large, with rainfall energy dominating the picture.

In this connection it should be pointed out that the fairly recent appreciation of energy in the erosions process, coming at a time when reliable data were scarce, has led to the widespread quotation of inaccuracies. It is easy to show that there is an error of a factor of about 100 in estimates such as 'the energy output of rain at 2 in. per hour on one acre is equal to that of six 40-horse power tractors'. However the literature already abounds with such statements (the present author was guilty in an early publication (HUDSON 1957), and since it has been made almost official by inclusion in the *USDA Yearbook* of 1955 (page 127) it will no doubt continue as a popular misconception.

Another approach to the part played by splash erosion and the role of surface runoff comes from Ellison's concept that erosion may be divided into detachment, or the tearing loose of small particles from their moorings in the soil mass, followed by their transportation, and finally deposition. It follows that the principle effect of raindrops is to detach soil,

Plate 4.1 High-speed photograph shows that the action of a falling raindrop is like a bomb striking soil surface (USDA)

while the principle effect of surface flow is the transportation of the detached soil. It is also clear that under conditions of heavy rain the detaching action of the raindrop splash is by far the most important part of the process. This was demonstrated in a simple field experiment (HUDSON 1957). Two small plots of bare soil, 1.5 m wide and 27.5 m long, received identical soil treatment (weeding and digging), but over one plot there were suspended two layers of mosquito gauze (a fine-gauge wire mesh) which allowed the rain to pass through, but broke up the rapidly falling raindrops into a spray of fine drops at lower velocity. On the second plot the bare soil was exposed to the rain (plate 4.4). The soil loss from the gauze plot, where splash erosion was eliminated, was reduced to one hundredth of the soil loss from the unprotected plot. The details of this experiment are given in section 11.4.

Raindrop impact has other important effects as well as particle detachment. The detached particles lead to sealing of the soil surface and hence to lower infiltration and increased surface runoff. More important,

Table 4.1 Kinetic energy of rain and runoff

Kinetic energy = $\frac{1}{2}$ × mass × (velocity)^2

	Rain	*Run-off*
Mass	Assume the mass of falling rain is R	Assuming 25% runoff, mass of runoff is $\frac{R}{4}$
Velocity	Assume terminal velocity of 8 m/sec	Assume speed of surface flow of 1 m/sec
Kinetic energy	$\frac{1}{2} \times R \times (8)^2 = 32R$	$\frac{1}{2} \times \frac{R}{4} \times (1)^2 = \frac{R}{8}$

The rain thus has 256 times more kinetic energy than the surface run-off.

the energy causes turbulence in the runoff, thus greatly increasing its capacity to scour and to transport soil particles. This effect has recently been studied by QUANSAH (1981) and KINNELL (1991).

4.2.2 *Erosion and energy*

There is a great deal of experimental evidence to suggest a link between erosive power and the mass and velocity of falling drops. In the early days of erosion research ELLISON (1944) in a laboratory experiment measured splash erosion for various combinations of drop size, velocity, and intensity. Analysis of his results gave the expression

$$S \propto V^{4.33} \times D^{1.07} \times I^{0.65}$$

where S is the grams of soil splashed in 30 minutes, V is the drop velocity in feet per second, D is drop diameter in mm, I is the rainfall rate in inches per hour.

A similar expression was obtained by BISAL (1960) who from similar laboratory experiments suggests $G = KDV^{1.4}$, where G is the weight of soil splashed in grams, K is a constant for the soil type, D is drop diameter in mm, and V is impact velocity in metres per second.

If erosive power is thus linked with combination of drop mass and drop velocity, it is not surprising that it has also been associated with kinetic energy. Indeed MIHARA (1951) found that splash erosion is directly correlated with kinetic energy, and FREE (1960) suggested that the relationship which best fits his experimental results was

for sand Splash Erosion $\propto$ (Kinetic Energy)$^{0.9}$

for soil Splash Erosion $\propto$ (Kinetic Energy)$^{1.46}$

ROSE (1960) challenged the assumption that results such as these prove that splash erosion depends uniquely on the kinetic energy of rain, and showed that if a relationship exists between erosion and energy, then an equally valid, though different, relationship will exist between erosion and momentum, or any other function of mass and velocity. Since mass occurs in the same form in the formula for both momentum and energy, it is necessary to vary velocity in order to solve the problem of whether energy or momentum is the better index of erosivity and this Rose did. After detailed experimental work, Rose concluded that the rate of detachment depends more closely on momentum than energy. However, it has been shown that for natural rain the relationships between intensity and either momentum or kinetic energy are of similar form (HUDSON 1965), so the issue of whether momentum should be used instead of energy is not important.

Several experimental studies have established correlations between soil splash and intensity (BUBENZER and JONES 1971, MOLDENHAUER and LONG 1964, MEYER 1981). The experimental evidence therefore suggests that both intensity and energy are closely linked with erosivity.

4.3 Estimating erosivity from rainfall data

There will obviously be some degree of correlation between erosive power and the amount of rainfall. A number of correlations have been established in localized conditions, for example in Malaysia MORGAN (1974) established a correlation between erosivity and the ten-year daily rainfall amounts. In Zimbabwe, ELWELL and STOCKING (1975) obtained reasonable agreement between erosivity and rainfall amount, based on the concept of selecting only rainfalls within defined limitations of amount and duration. However, the large variations in rainfall, both temporal and spatial, mean that this approach can only yield empirical results of limited local application.

The attraction of using readily available data on amounts of rainfall has also led to a search for a method which will allow world-wide application. The *Fournier Index* was intended as a guide to the potential for geological erosion on a large scale, and was based on correlating rainfall with sediment load measured in the major rivers of Africa (FOURNIER 1960). A modified version was introduced for the FAO study of *Soil Degradation* (ARNOLDUS 1980), and a first approximation of a world-wide map of erosivity using this modified Fournier index has been produced (KINGU 1980). From the crude nature of the input, such models can only give a first approximation of erosivity, and more accurate estimations must depend on more detailed inputs.

4.3.1 The EI index

The laboratory experiments by Ellison, Bisal, Rose and others certainly suggested that erosivity is related to energy, but it also had to be verified that the hypothesis holds for soil losses in the field caused by natural rain.

The confirmation came from the work of WISCHMEIER (1955) in the United States. At the Erosion Research Data Processing Laboratory at Purdue University, Wischmeier put on punch-cards all the data from the experimental plots of 35 conservation experiment stations; a total of 8 250 plot years records. Using machine processing of punch-cards carrying data, and computers to perform the calculations, a search was made for parameters which could be correlated with the recorded erosion. The results, summarized by WISCHMEIER *et al* (1958) showed that the factor most closely related to erosion was the kinetic energy of the rain. However, there was still considerable unexplained variation, so multiple regressions were also tested. The best estimator of soil loss was found to be a compound parameter, the product of the kinetic energy of the storm and the 30-minute intensity. This latter term requires explanation. It is the greatest average intensity experienced in any 30-minute period during a storm. It is computed (figure 4.3) from recording rain gauge charts by locating the greatest amount of rain which falls in any 30-minute period, and then doubling this amount to get the same dimensions as intensity, i.e. rainfall per hour.

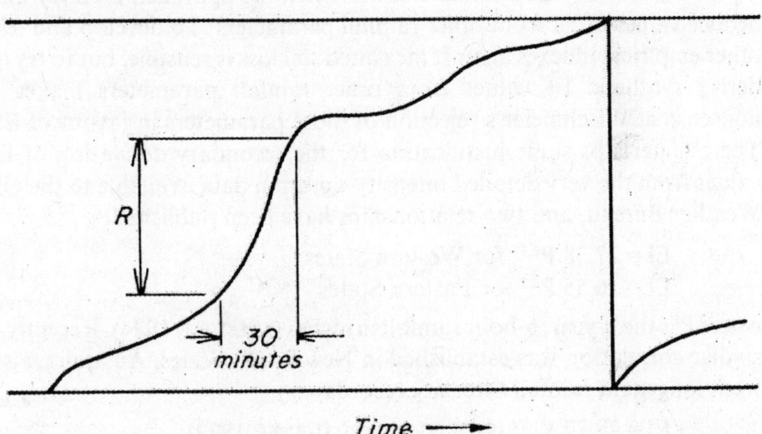

Figure 4.3 Method of obtaining the 30-minute intensity. The 30-minute period with the greatest amount of rain, *R*, is found from the rain gauge chart. Twice this amount is the 30-minute intensity

This product term gave an excellent correlation. Wischmeier found that it was in fact possible to improve the relationship even further by taking account of other factors, such as the soil moisture at the beginning of the storm, but these further refinements added so little to the accuracy that they were felt to be not worth the extra complication. This measure of erosivity is described as the *EI index*. It can be computed for individual storms, and the storm values can be summed over periods of time to give weekly, monthly, or annual values of erosivity.

The use of EI as an index of erosivity and its valuable application in the Universal Soil Loss Equation have led to over-enthusiastic extension beyond the purpose for which it was designed (WISCHMEIER 1976). One hazard is that although there is a high degree of correlation over the long term, there are large short-term variations. EI values for one year can range from 50% to 200% of the long-term average. The use of the EI index in the USLE is discussed in section 7.2.2.

The second problem is that attempts to apply the original index to tropical rainfall led to some excessively high estimates. The revised method (WISCHMEIER and SMITH 1978) allows a maximum value of energy as shown in figure 3.9, and a maximum of 63.5 mm/h for I_{30}. This reduces the errors when applying the method to tropical rainfall, but the fact that after 15 years the rules had to be substantially changed illustrates the folly of extrapolating an empirical formula into situations for which it was not intended.

Another equally unprofitable and unscientific approach is to try and deduce values of EI from other rainfall parameters. To develop and test other empirical indexes against measured soil loss is sensible, but to try to derive synthetic EI values from other rainfall parameters makes a nonsense of Wischmeier's rejection of those parameters in favour of EI. There is perhaps some justification for the secondary derivation of EI values from the very detailed intensity-duration data available to the US Weather Bureau, and two relationships have been published.

$$EI = 27.38\ P^{2.17} \text{ for Western States}$$

$$EI = 16.55\ P^{2.2} \text{ for Eastern States}$$

where P is the 2-year, 6-hour rainfall in inches (ATESHIAN 1974). Recently a similar correlation was established in New South Wales, Australia, also from long-term rainfall intensity records

$$EI = 29.22\ P^{1.89}$$ (ROSEWELL and TURNER 1992)

4.3.2 The KE>25 Index and the AI_m Index

In several studies in the tropics the EI index has been found to be less effective as a measure of erosivity than might be expected from

Wischmeier's studies in America. This has led to a search for alternative methods. One, developed by the present author in Africa, is based on the fact that little erosion takes place at low intensities.

At low intensity, rain is composed mainly of small drops, falling with low velocity, and hence low energy. Even if a little splash erosion does occur there is usually no runoff to carry away the splashed particles. Experiments were carried out to see whether there is a recognizable point below which rain is non-erosive. Tests with Ellison splash cups (described in section 4.4) showed that although there is variation from one storm to another the intensity of 25 mm per hour can be taken as a practical threshold separating erosive and non-erosive rain. An erosivity index, written as KE>25, consisting of the total kinetic energy of all the rain falling at more than 25 mm per hour was found to give an excellent correlation with soil loss, as explained in section 4.4.

The concept has been modified for temperate use by MORGAN (1977) using a lower threshold value of 10 mm/h. The idea of using threshold values for both intensity and amount was also applied by ELWELL and STOCKING (1975), and it is built into the latest methods of calculating EI by ignoring showers of less than 0.5 inch and separated from other rain periods by more than 6 hours unless 0.25 inch fell in 15 minutes (WISCHMEIER and SMITH 1978).

In Nigeria, Lal tested EI, KE>25, and other possible parameters, and found the best correlation of soil loss from small plots with a new index rather similar to EI. This is AI_m, where A is the amount of rain, and I_m the maximum intensity over a 7.5 minute period. As with previous studies, the correlation could be improved by the addition of other factors, particularly a measure of wind speed which affects impact velocity and hence energy. However, transient phenomena, such as wind, cannot be included in an index intended for long term predictions (LAL 1976).

The tendency to use empirical indexes should be treated with reserve. Wischmeier had 10 000 plot years of data extending over 22 years and this solid database was used effectively to generate the empirical index EI. The present author had 13 years results, giving 2 500 plot years to test against KE>25. In Australia, the New South Wales Conservation Service now has a data base of 4 500 plot-years from a research programme extending over 40 years. Lal in Nigeria, and Roose in West Africa had smaller but significant bases (ROOSE 1980). But much time and effort have been wasted on trying to generate or test empirical equations against the results of a handful of plots for one or two seasons. Even more fruitless is the too popular pastime of comparing one empirical index against another without any database.

Another common error is to assert that one index is 'better' than others without a proper comparison. Several studies ignore the fact that alternative indexes give such high coefficients of correlation that the differences are irrelevant. An important point which reinforces the danger of empirical indexes is that all of us who tested EI against measured soil loss, or compared EI with other indexes, before the 1978 revisions to the USLE calculation procedures for EI, should do the calculations again using the new procedures.

4.3.3 Calculation procedures

The calculations for EI and KE>25 are similar, and shown in table 4.2. For both methods is it necessary to know the amount of rain which falls at specified rates of intensity. The amount of rainfall in each class of intensity is multiplied by the appropriate energy value, and the energy is totalled for the whole storm. In the EI method this total energy is then multiplied by the 30-minute intensity to give the erosivity value. In the KE>25 method the first line is omitted from the calculation and the index is simply the total energy of the remainder. The fact that the I_{30} value is not required may be a distinct advantage in countries with limited rainfall data, for it can only be obtained from autographic rain-gauge charts and is not ordinarily recorded by meteorological stations.

4.4 Splash erosion and wash erosion

Laboratory studies have usually compared estimates of erosivity with erosion measured as the amount of sand splashed out of small containers, whereas field experiments record the amount of soil washed into collection tanks at the lower edge of experimental plots. If an index of erosivity is set up to be an estimator of erosion it should be tested to see that it works as well for real field erosion as for small-scale laboratory experiments. The KE>25 index was so tested in a series of experiments starting with the special case of splash of sand and progressively adding the other variables.

First, splash erosion was measured alone using Ellison-type splash cups (plate 4.2). These were first used by ELLISON (1944) and have been used subsequently by many research workers because they offer a simple but precise and reproducible measurement of splash. Brass cylinders 77 mm in diameter and 50 mm deep have a wire mesh soldered on one end to form a porous base. A thin layer of cotton wool rests on the wire and the containers are filled with fine sand. When placed in a shallow depth of water the cups are maintained at a constant moisture status of near saturation. The cups are oven-dried and weighed before and after exposure to rainfall and, since the soil conditions have been standardized,

Table 4.2 Sample calculations of erosivity from rainfall records

(a) Using the EI method

1	2	3	4
Intensity (mm/h)	Amount (mm)	Energy[1] ($J/m^2/mm$)	Total (Col 2 × Col 3)
0–25	30	22.0	660
25–50	25	26.0	650
50–75	15	28.0	420
>75	10	28.3	283
			2013J/m²

I_{30} is derived from raingauge charts as shown in figure 4.3. Using a typical value of say 20 mm/h

$$EI = \frac{E \times I_{30}}{1000} = \frac{2013 \times 20}{1000} = 40.26 \text{ erosivity units } (^2)$$

Note (1) Calculated from $E = 11.9 + 8.7 \log I$ or figure 3.9

Note (2) In English units, and in the metric version in *Agriculture Handbook 537*, EI was divided by 100 to give convenient units, and in this example 1000 has been similarly used.

Note (3) The greatest care is required when handling EI in metric or SI units. The metric conversions on p 66 of early versions of *Agriculture Handbook 537* are inverted. The supplement to 537 of Jan 1981 differs from the proposals by FOSTER *et al* 1981. USA preference is for megajoules per hectare, but the natural SI unit of J/m^2 is generally accepted in Europe.

(b) **Using the KE>25 method**

1	2	3	4
Intensity (mm/h)	Amount (mm)	Energy[4] ($J/m^2/mm$)	Total (Col 2 × Col 3)
0–25	30	—	—
25–50	25	25.2	630
50–75	15	27.8	417
>75	10	29.0	290
	Total 80		1337J/m²

Note (4) Calculated from $E = 30 - \frac{125}{I}$ or from figure 3.9

the weight of sand splashed out gives a relative measurement of the 'splashability' of the rain. With developments in experimental technique (BISAL 1950, HUDSON 1965), this method gives a simple but precise measurement. In the experiments to test KE>25 groups of splash cups were exposed to every storm for two years, and the correlation between sand splash and KE>25 was very good at $r = 0.96$.

In the next stage surface runoff was introduced and soil used instead of sand in soil pans after the style of those used by FREE (1952). The pans were

Plate 4.2 Ellison-type splash cups measured the amount of a standard sand which is splashed out of the containers during a storm

1 metre long, 300 mm wide, and 100 mm deep, and set on a 5% slope (plate 4.3). They were filled with a sieved clay loam soil which was maintained at constant moisture status by connecting each tray to a constant-head reservoir. The soil washed off the pan was collected in a trough at the lower edge. Again after two years of results the correlation between KE > 25 and the measured soil loss was still remarkably good at $r = 0.92$.

The next stage was to use a field scale plot 27.5 metres long by 1.5 metres wide on a 5% slope (plate 4.4). The soil was undisturbed except for the minimum amount of weeding to keep the surface free of vegetation, and the soil moisture was not controlled. The soil loss was measured by catching the runoff in tanks at the bottom of the plot. In spite of having now re-introduced all the variables which occur in nature, such as variations in the soil moisture when the storm starts, the correlation was still $r = 0.94$.

It is clear then that reliable estimates of the erosive power can be calculated from rainfall records, and in the EI index and the KE > 25 index

Plate 4.3 Soil trays for measuring splash and wash erosion. The sixth tray in the background serves as a constant-head tank, maintaining constant moisture level in the other five trays. The porous tiles in the false floor are seen in the empty tray on the left

we have alternative methods, slightly different in approach but both tested and proved to be suitable for the purpose and efficient in operation.

4.5 Applications of an index of erosivity

The ability to assess numerically the erosive power of rainfall has two main applications. In practical soil conservation it helps to improve the design of conservation works, and in research it helps to increase our knowledge and understanding of erosion. Some examples will illustrate these uses.

4.5.1 An aid to design

The soil conservation techniques used to combat erosion must be designed to do their job efficiently. If they are under-designed they will fail; if they are overdesigned, i.e. unnecessarily large or complicated, this is equally undesirable. The right balance can only be achieved if we know fairly accurately the severity of the problem which has to be faced, and a

Plate 4.4 The field plot whose soil loss was used to test the estimates of erosivity from rainfall records

quantitative measure of erosivity makes it possible to identify the basic erosion risk in different areas. For example, if it is known that the erosivity of the rain in the northern part of a country is two and half times as great as in the south, then it would be sensible to design terraces according to different specifications in the two regions.

In any country with sufficiently detailed rainfall data, maps of erosivity could be produced, as has been done in the United States of America. The total annual erosivity can be drawn on a map as in figure 4.4 and used to predict annual erosion losses, section 7.2.2. Information can also be tabulated on how erosivity varies during the year at any given location (figure 4.5) and this can be used for the detailed planning of conservation farming systems.

Another practical demonstration can be taken from an experience in Africa. When carrying out gully reclamation works which include the establishment of vegetation it is essential to have some rain to encourage growth, but highly erosive storms are likely to wash out the newly planted material. In tropical climates with a concentrated rainy season this means choosing either the beginning or the end of the rains. Figure 4.6 shows

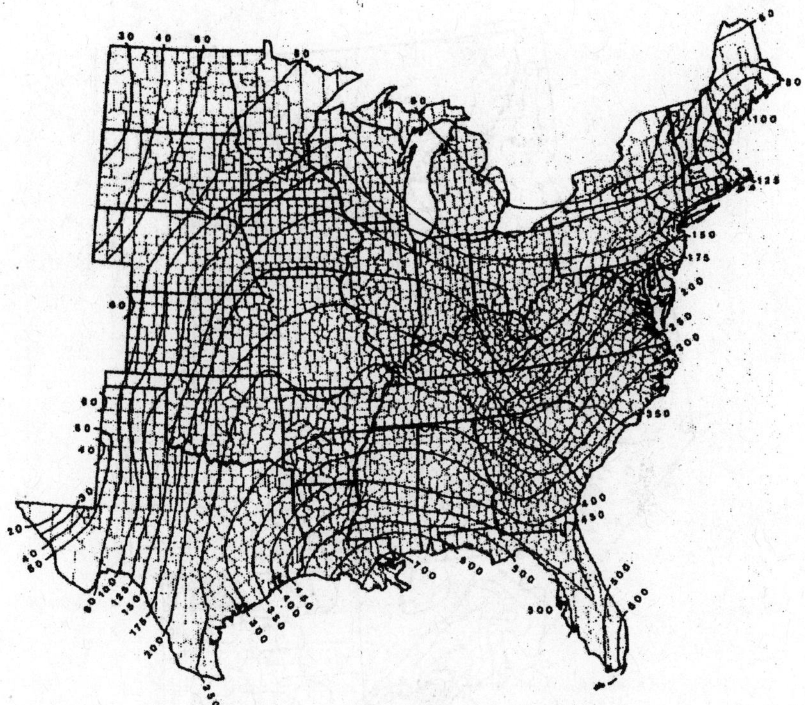

Figure 4.4 Map of average annual values of the rainfall erosion index of eastern United States (*Agricultural Handbook 703*, 1994)

how a monthly plot of values of erosivity gave the answer. Bulawayo has low values of erosivity at the beginning of the rainy season in November and December, but high values in February, so the beginning of the season is evidently the best time for establishing anti-erosion vegetation. On the other hand at Mutari the rainy season opens with high erosivity storms, and the better time would be in February and March when the erosivity falls away.

Any such calculations of erosivity must be based on average rainfall figures and a particular month or year could differ considerably from the average. The fluctuations of erosivity about the average will follow fluctuations in total rainfall, and so more reliable estimates of erosivity can be made where the rainfall is reliable, than where it is erratic. The length and accuracy of rainfall records will also affect the accuracy of estimates of erosivity.

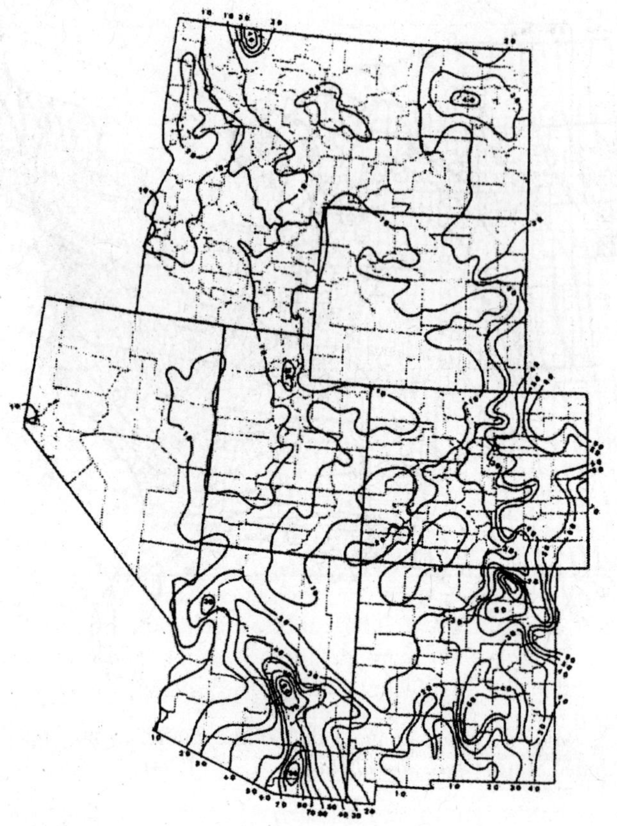

Figure 4.5 Map of average annual values of the rainfall erosion index for the western United States (*Agricultural Handbook 703*, 1994)

4.5.2 An aid to research

Soil erosion research uses either natural rainfall or man-made simulated rain and in both cases a quantitative scale of erosivity is a valuable asset. One of the drawbacks of using field plot experiments to measure the runoff and soil loss under natural rainfall is that it takes so long to get reliable results. Experiments concerned with rotations are particularly difficult. A plot will have its particular treatment one year and the soil loss is recorded. The next year it moves to the next phase of the rotation and a different soil loss is recorded. The difficulty is to separate how much of this difference is due to the different cropping treatments and how much due to the different rainfall. Neither replications nor statistical

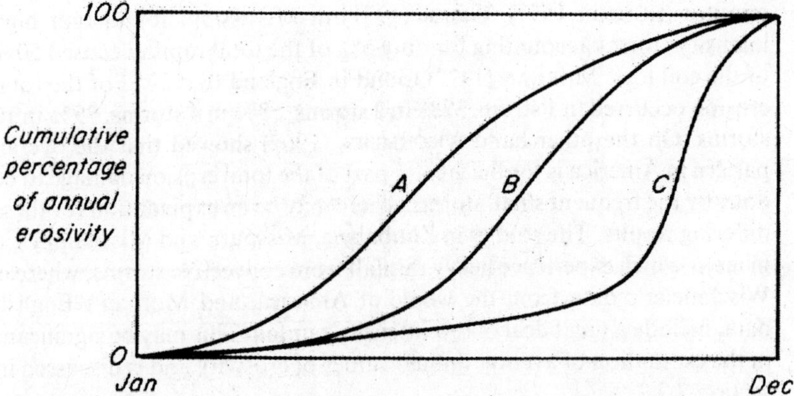

Figure 4.6 Cumulative erosivity during the year. Curve *A* indicates fairly even distribution of erosivity, curve *B* mainly summer rains, and curve *C* heavy autumn rains

techniques can completely overcome this problem, but if it is possible to numerically evaluate the erosivity, then allowance can be made for the variations from year to year. This enables much more information to be obtained in limited time from field experiments. The same argument applies to the variations during one season. Suppose the measured soil loss is found to be different in the early part of the season from that later on. The question is 'does the difference arise from the change in crop growth or is the rain different during these two periods?' There is no answer without a means of measuring erosivity, but to measure the effect of crop growth on erosion is simple when the erosivity of each storm can be compared.

4.5.3 *Is erosion cataclysmic or slow attrition?*

An important question which occurs wherever there is a serious erosion problem is 'what kind of rain does the greatest amount of damage in the long term? Is it the small storms which only cause a little erosion but happen very often, or is it the tremendous storm which only happens occasionally but does more damage than all the little storms?' The experimental evidence differs widely. LITTLE (1940) quotes one extreme; a single storm at the experiment station at Bethany, Missouri, caused 50% of the total erosion in five years. At the experiment station at Mazowe, Zimbabwe, it was found that in almost all seasons more than half of the total amount of erosion occurred in the one or two heavier storms of the year. In one case three-quarters of the yearly soil loss took place in ten minutes (HUDSON 1957). GREER (1971) in Mississippi found that high intensity storms accounting for only 6% of the total rainfall caused 50% of the soil loss. MORGAN (1977) found in England that 17% of the total erosion occurred in 1 storm, 32% in 2 storms, 58% in 4 storms, 99% in 10 storms. On the other hand WISCHMEIER (1962) showed that the overall pattern in America is for the greater part of the total erosion damage to be done by the frequent small storms. There may be an explanation for these differing results. The studies in Zimbabwe, Missouri, and Mississippi are in areas which experience heavy rainfall from convective storms, whereas Wischmeier's data from the whole of America, and Morgan's English data, include a great deal of low intensity rainfall. This may be significant in the estimation of average annual values of erosivity and is discussed in section 7.2.2.

4.5.4 Temperate and tropical rainfall

Probably the most important application of the measurement of erosivity is that it allows a simple explanation of why soil erosion is a serious problem in tropical and sub-tropical countries, but much less so in temperate climates. It was shown in section 4.3.2 that rain falling at low intensities is non-erosive and the threshold level of intensity at which it becomes erosive is about 25 mm per hour. Figure 4.7 shows the kind of distribution obtained by plotting the amount of rain falling at different intensities. The vital difference is that in temperate rainfall something like 95% of the rain falls at low non-erosive intensities, i.e. only 5% is heavy enough to cause erosion, whereas in the case of tropical rainfall only 60% falls at intensities less than 25 mm/hour, and the remaining 40% contributes to soil erosion.

These are two further factors contributing to the difference. On the whole it is probable that the total quantity of rain will be greater in tropical climates so a comparison of the amount of erosive rain might be:

> *Temperate climate* with 5% of the rain being erosive,
> 5% of say 750 mm of annual rain = 37.5 mm of erosive rain
> *Tropical climate* with 40% of the rain being erosive,
> 40% of say 1500 mm of annual rain = 600 mm of erosive rain.

Another difference is that the average intensity of the erosive rain is higher in the tropical distribution – perhaps about 60 mm per hour as in figure 4.8 compared with about 35 mm per hour for the temperate rain. It was shown in figure 3.9 that the kinetic energy per mm of rain increases as the intensity increases, so to calculate the total erosivity we would use a higher figure, perhaps 28 joules per square metre per mm for tropical rain,

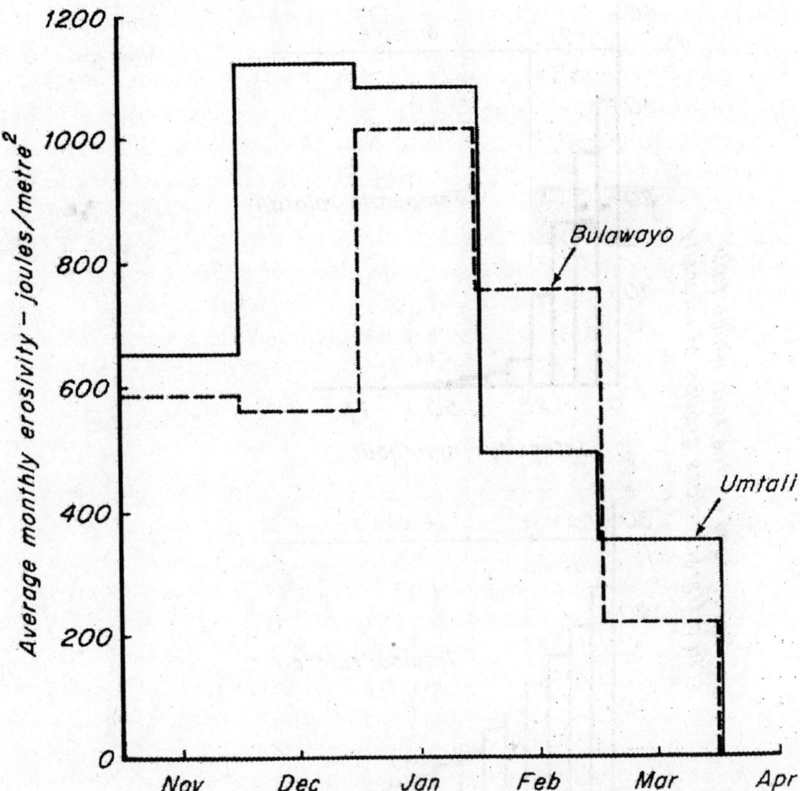

Figure 4.7 Average monthly values of rainfall erosivity for two towns in Zimbabwe

compared with perhaps 24 for temperate rainfall. Using the same figures as in the example above, the annual erosivity values would be

Temperate rain $37.5 \times 24 = 900$ joules/metre^2
Tropical rain $600 \times 28 = 14400$ joules/metre^2.

The erosive power of the tropical rain is therefore 16 times greater than that of the temperate rain. The actual numbers used in this example are not important. What is **important** is that this simple calculation shows how tropical rainfall is likely to have a very much greater power to cause erosion, and this is why erosion is so much more serious in countries with tropical or sub-tropical rainfall.

However, the amount of erosion which takes place depends not only on

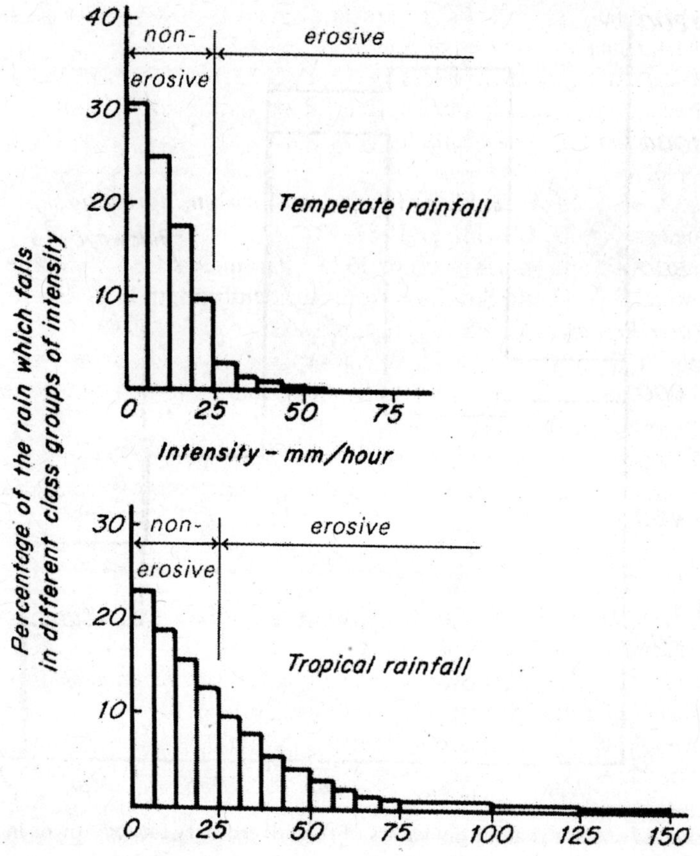

Figure 4.8 The distribution of rainfall at different intensities for tropical and temperate rainfall

the rain but also on the soil surface on which the rain falls. This aspect, the erodibility, will be considered in chapter 5.

References

ARNOLDUS, H. M. J. 1980 An approximation of the rainfall factor in the USLE, in *Assessment of Erosion*, edited by de Boodt and Gabriels, Wiley, Chichester, Sussex

ATESHIAN, J. K. H. 1974 Estimation of Rainfall Erosion Index, *Journal of Irrigation and Drainage Division, American Society of Civil Engineers*, 100, IR3, 293–307

BISAL, F. 1960 The effect of raindrop size and impact velocity on sand splash *Canadian Journal of Soil Science*, 40, 242–245

BUBENZER, G. D. and B. A. JONES 1971 Drop Size and Impact Velocity Effects on the Detachment of Soils under Simulated Rainfall, *Transactions of American Society of Agricultural Engineers*, 14, 4, 625–628

ELLISON, W. D. 1944 Studies of Raindrop Erosion, *Agricultural Engineering*, 25, 131–136 and 181–182

ELWELL, H. A. and M. A. STOCKING 1975 Parameters for Estimating Annual Runoff and Soil Loss from Agricultural Lands in Rhodesia, *Water Resources Research*, 11, 4, 601–605

FOSTER, G. R., D. K. MCCOOL, K. G. RENARD and W. C. MOLDENHAUER 1981 Conversion of USLE to SI units, *Journal of Soil and Water Conservation*, 36, 6, 355–359

FOURNIER, F. 1960 *Climat et èrosion, La relation entre l'érosion du sol par l'eau et les précipitations atmosphériques*, Presses Universitaires de France, Paris

FREE, G. R. 1952 Soil movement by raindrops, *Agricultural Engineering*, 33, 491–494, 496

FREE, G. R. 1960 Erosion characteristics of rainfall, *Agricultural Engineering* 41, 7, 447–449, 455

GREER, J. D. 1971 Effects of excessive-rate rainstorms on erosion, *Journal of Soil and Water Conservation*, 26, 196–197

HUDSON, N. W. 1957 Erosion Control Research: Progress Report on Experiments at Henderson Research Station 1953–1956, *Rhodesian Agricultural Journal*, 54, 4, 297–323

HUDSON, N. W. 1965 *The influence of rainfall on the mechanics of soil erosion*, MSc Thesis, University of Cape Town

KINGU, P. A. 1980 *World Map of Erosivity*, MSc Thesis MS/80/190, National College of Agricultural Engineering

KINNELL, P. I. A. 1991 The effect of flow depth on erosion by raindrops impacting shallow flow, *Transactions of the Association of Agricultural Engineers*, 34, 1, 161–168

LAL, R. 1976 Soil Erosion on Alfisols in Western Nigeria; III, Effects of Rainfall Characteristics, *Geoderma*, 16, 389–401

LITTLE, J. M. 1940 *Erosional Topography and Erosion*, Carlisle, San Francisco

MEYER, L. D. 1981 How rain intensity affects interrill erosion, *Transactions of the American Society of Agricultural Engineers*, 24, 6, 1472–1475

MIHARA, Y. 1951 Raindrops and Soil Erosion, *Bulletin of Natural Institute of Agricultural Science Series A, 1*

MOLDENHAUER, W. C. and D. C. LONG 1964 Influence of rainfall energy on soil loss and infiltration rates 1, Effect over a range of textures. *Proceedings Soil Science Society of America*, 28, 6, 813–817

MORGAN, R. P. C. 1974 Estimating regional variations in soil erosion hazard in peninsular Malaysia, *Malaysia Nature Journal*, 28, 94–106

MORGAN, R. P. C. 1977 Soil Erosion in the United Kingdom: field studies in the Silsoe area 1973–75, *National College of Agricultural Engineering, Occasional Paper 4*

QUANSAH, C. 1981 The effect of soil type, slope, rain intensity and their interactions on splash detachment and transport, *Journal of Soil Science*, 32, 2, 215–224

ROOSE, E. J. 1980 Approach to the definition of rain erosivity and soil erodibility in West Africa, in *Assessment of Erosion*, edited by de Boodt and Gabriels, Wiley, Chichester, Sussex

ROSE, C. W. 1960 Soil detachment caused by rainfall, *Soil Science*, 89, 1, 28–35

ROSEWELL, C. J. and J. B. TURNER 1992 Rainfall Erosivity in New South Wales, *Technical Report 20*, Department of Conservation and Land Management, NSW

USDA, 1955 *Water: Yearbook of Agriculture 1955*, United States Department of Agriculture, Washington DC

USDA-ARS 1994 Predicting Soil Erosion by Water: A Guide to Conservation Planning with the Revised Universal Soil Loss Equation (RUSLE), *Agriculture Handbook 703*, United States Department of Agriculture, Washington DC

WISCHMEIER, W. H. 1955 Punch cards record runoff and soil loss data, *Agricultural Engineering*, 36, 664–666

WISCHMEIER, W. H. 1962 Storms and soil conservation, *Journal of Soil and Water Conservation*, 17, 2, 55–59

WISCHMEIER, W. H. 1976 Use and misuse of the Universal Soil Loss Equation, *Journal of Soil and Water Conservation*, 31, 1, 5–9

WISCHMEIER, W. H. and D. D. SMITH 1978 Predicting rainfall erosion losses, *Agriculture Handbook 537*, United States Department of Agriculture, Washington DC

WISCHMEIER, W. H., D. D. SMITH and R. E. UHLAND 1958 Evaluation of Factors in the Soil-Loss Equation, *Agricultural Engineering*, 39, 8, 458

5

The erodibility of soil

5.1 Definitions

The *erodibility of soil* is its vulnerability or susceptibility to erosion, that is the reciprocal of its resistance to erosion. A soil with a high erodibility will suffer more erosion than a soil with low erodibility if both are exposed to the same rainfall. Whereas the *erosivity of rainfall* was shown in chapter 4 to be a fairly straightforward measure of the rain's physical properties, the assessment of erodibility is more complicated because it depends upon many variables. There are alternative uses of the term *erodibility*. In the widest sense it can be used to include all the variables which effect erosion except the erosivity of rain, as in the first line of figure 2.2. It is also used in the narrower sense as a measure only of the effect of physical characteristics of the soil, as in the erodibility index *K* of the Universal Soil Loss Equation, where the effects of land management and crop management are assessed separately, as shown in the last line of figure 2.2.

5.2 Factors influencing erodibility

In this chapter we will look at the three broad groups of factors which affect erodibility. First, there are the physical features of the soil, including the chemical and physical composition. Second, there are the topographic features, such as the slope of the land. Third, there is the management of the land, i.e. how it is used.

5.2.1 Soil physical characteristics

For many years soil scientists have attempted to relate the vulnerability of soil to physical properties which can be measured in the laboratory or the field. Pioneer work in America in the 1930s attempted to explain the result of early field erosion experiments in terms of physical and chemical properties (LUTZ 1934). A host of studies have followed, and only a selection will be reviewed to illustrate the principles which have been used.

Laboratory analysis An obvious factor to test is the texture or mechanical composition, and BOUYOUCOS (1935) suggested as an index of erodibility the ratio

$$\frac{\%\ \text{sand} + \%\ \text{silt}}{\%\ \text{clay}}$$

Since then many variations on this theme have confirmed the basic feature that sand and silt tend to increase erodibility, while clay decreases it. BARNETT and ROGERS (1966) suggested an index similar to that of BOUYOUCOS. The index of WISCHMEIER *et al* (1971) has for its main variables the components (% silt + % very fine sand) and % sand greater than 0.1 mm.

Another logical assumption is that resistance to water erosion will be linked to the degree of aggregation into soil particles, and the stability of the particles, so many workers have devised methods for quantifying this. MIDDLETON *et al* (1934) and HAMILTON (1977) use a measure of dispersion in water, while HENIN (1963) and VOZNESENSKY and ARTSRUUI (1940) use chemical dispersion, and YODER (1936) and BRYAN (1968) chose to measure the water-stable aggregates. More recent studies tend to confirm the importance of water stable aggregates as one of the more important physical characteristics. ELWELL (1986) suggests that the most useful single parameter is the mean weight diameter of water-stable aggregates, but since this is time consuming to measure he suggests that the proportion of water-stable aggregates of diameter more than 2 mm is an acceptable alternative and much easier to measure. PAEZ and PLA (1987) suggest that indices based on the assessment of water-stable aggregates by the usual technique of wet-sieving tend to underestimate the erosion risk, particularly in soils of medium to high erodibility, and suggest that aggregate stability is better measured under the impact of raindrops, particularly if the technique can also evaluate the sealing effect of the fine particles. DANGLER *et al* (1987) studied a huge number of physical characteristics which can be quantified in the laboratory and concluded, like Elwell, that the reliability of estimates of erodibility was not seriously reduced by using only parameters which may be easily and simply measured, and suggested that the two most important factors are the percentage of unstable aggregates and suspension percentage. They also showed that any estimates of erodibility are to some extent dependent upon the parent material with significant differences between residual oxisols and volcanic ash soils.

Since aggregate stability is not easily measured, an alternative is to substitute some more easily measured feature which can be expected to have a similar effect. Examples are PEELE (1937) using percolation, NAMBA (1952) using infiltration, or WISCHMEIER *et al* (1971) who include organic matter, structure, and permeability, in their nomograph shown in figure 5.1. LAL (1988) presents a table of the more important indexes based on parameters that can be measured in the laboratory.

In section 2.2.5 we discussed how resistance to erosion can depend on the form of erosion. It could be that the importance of these measurable physical characteristics is different for the different erosion processes. For example in the high-energy detachment of particles and the breakdown of aggregates, resistance may depend mainly on the strength of aggregate stability, while in the transport dominated situation resistance to flow may depend more on the size of aggregates.

Laboratory rainfall studies Other research workers have wished to reproduce more accurately the real conditions of rainfall erosion by subjecting soil samples to the action of falling water drops. The simplest variation is single water drops falling from a burette, as used by MCCALLA (1944) and RAI *et al* (1954). Multiple droppers to simulate the effect of rain were used by PEREIRA (1956) and MOLDENHAUER (1965) and a detailed study of indices of soil erodibility using different simulators was carried out by BRYAN and DE PLOEY (1983). They found that minor variations in the design of rainfall simulators are only significant when soils have a very high sand content or poor aggregation, and that for other soils, laboratory testing under simulated rainfall appears to produce valid, comparable results encouraging confidence in the value of the technique. They also reaffirm Bryan's earlier conclusion 'that most indices are unreliable when applied to more than a very limited range of soils'. They confirmed other studies indicating the importance of soil aggregation with a preference for the percentage weight of water-stable aggregates more than 0.5 mm diameter.

However, all laboratory studies have to be carried out on soil samples, with the attendant uncertainty as to whether the process of removing the sample may have changed the very properties which are being measured. Also laboratory analysis of particle size distribution is usually done after shaking a sample with a dispersing agent. Several workers now suggest that this is a more aggressive disaggregation than takes place in the field under raindrop impact and surface flow (LOCH and ROSEWELL 1992).

Field studies Field tests of erodibility on undisturbed soil have been

Figure 5.1 The soil erodibility nomograph (WISCHMEIER *et al* 1971). The units of *K* are for the original units of *R* in foot-tons/acre-inch. For conversion to SI units divide *K* values of this nomograph by 7.49 (*Agricultural Handbook 703*, 1994)

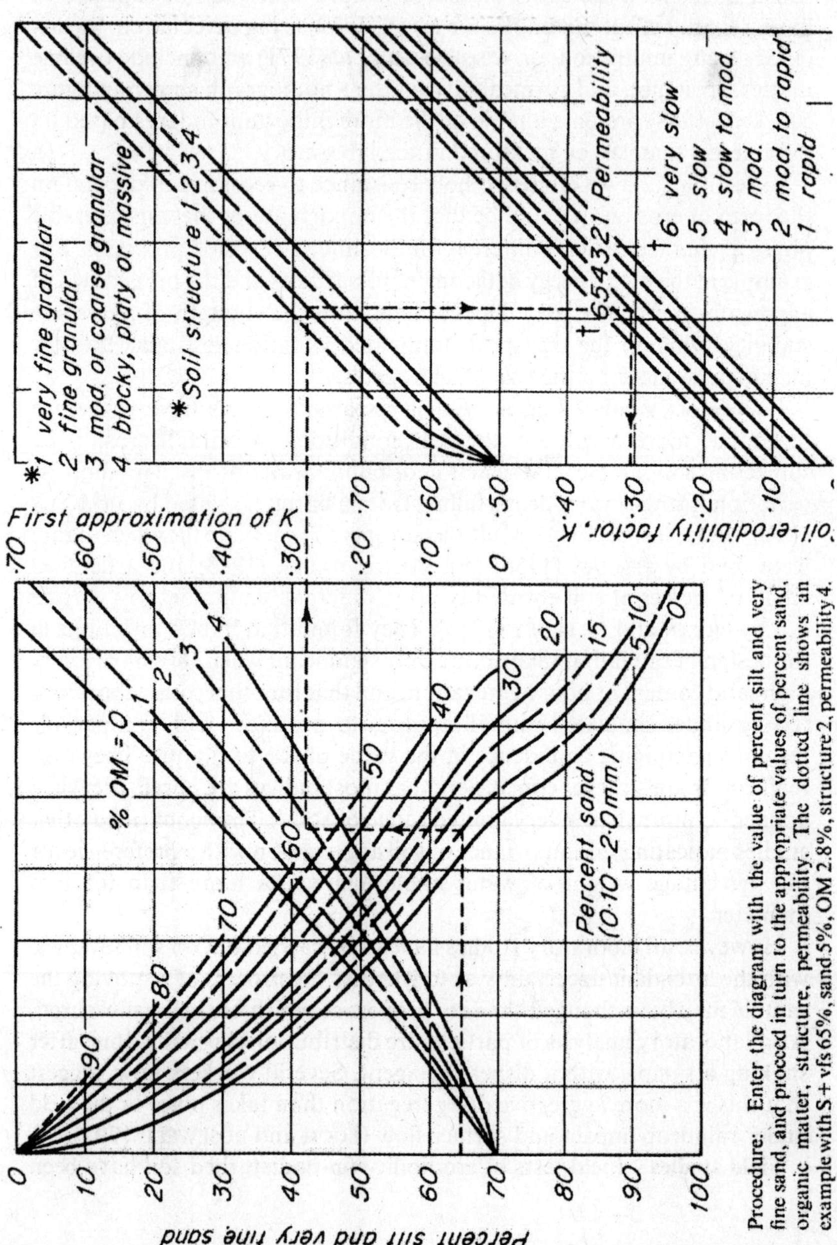

Procedure: Enter the diagram with the value of percent silt and very fine sand, and proceed in turn to the appropriate values of percent sand, organic matter, structure, permeability. The dotted line shows an example with S + vfs 65%, sand 5%, OM 2.8%, structure 2, permeability 4.

carried out under natural rainfall and using artificial rain-making devices. Few studies are recorded of experiments deliberately set up to test the erodibility of different soils under natural rainfall because of the physical difficulties of establishing such experiments, and the problem of eliminating variations caused by rainfall at different sites. Even the massive data-bank of the USDA-SCS of 4000 plot-years produced only seven assessments of erodibility based on similar conditions under natural rainfall (WISCHMEIER *et al* 1958), and these were the by-product of experiments set up for other reasons. Purpose-built experiments on erodibility under natural rain were carried out by ROOSE (1977) and OLSON and WISCHMEIER (1963) but the main thrust has come, and will continue to come, from studies using artificial rainfall simulators. The first of such experiments used a simple portable device (ADAMS *et al* 1957), but since then the rainfall simulator has become a much more sophisticated and powerful research tool as discussed in section 8.4. Some notable field studies of erodibility using rainfall simulators are MEYER and MCCUNE (1958) at Purdue University, BARNETT and ROGERS (1966) in Georgia, BRYAN in England (1968), DANGLER *et al* in Hawaii (1975), KHATRI CHETRI and PAINTER in New Zealand (1971), and ARMSTRONG in Australia (1990).

There will always be a place for empirical methods to deduce erodibility from other more easily measurable properties, but soils vary greatly in their origin and manner of formation, as well as their management. The greatest care should therefore be exercised in extrapolating empirical methods from one soil to another, and particularly between temperate and tropical soils. EL-SWAIFY *et al* (1982) have shown that tropical soils show extreme variability in erosion susceptibility, with K values varying from 0.06 to 0.48. VANELSLANDE *et al* (1984) found wide variations between measured values of 3 soils of 0.015, 0.04, and 0.04 which had corresponding values on the Wischmeier nomograph of 0.39, 0.025, and 0.018. All indirect methods of estimating erodibility should therefore be regarded as second-best substitutes for direct measurement from undisturbed soil.

5.2.2 *Variations in erodibility*

For convenience in making quantitative estimates of erosion, using models such as the USLE, we assume that the long-term average erodibility of soil can be represented by a single factor. However, there are two reasons why this is an approximation. First, we have raised the possibility that erodibility can vary with different erosion processes, and that it could interact with erosivity. The second is that it can change with time and with management practices. YOUNG *et al* (1990) have shown that there are seasonal variations in erodibility which are determined by seasonal variations in temperature and soil moisture. ELWELL (1986) has shown in Zimbabwe that erodibility can be increased by cultivations which can break down soil aggregates and by reducing the amount of organic matter in the soil. Equally, practices which increase the organic matter content, such as mulching or manuring, can reduce the erodibility. Changes with time have been recorded by VANELSLANDE *et al* (1984) on Nigerian soils and by ROOSE and PIOT (1984) in Burkina Faso. It is unlikely that trying to modify erodibility by changing the aggregation through chemical additives will be economic on a large scale. However, the treatment of small areas which are temporarily vulnerable to severe erosion, such as roadsides, is certainly possible and is discussed in section 13.2.

5.2.3 Topography

Expressed in simplest terms, steep land is more vulnerable to water erosion than flat land for the obvious reason that the erosive forces, splash, scour, and transport, all have a greater effect on steep slopes. This effect can be quantified at several scales.

At the macro-scale, i.e. regional or continental, BOSAZZA (1953) showed that in southern Africa the severity of the relief expressed in terms of the gradients of the major rivers is one of the factors correlating with severe geological erosion. Modern satellite imagery shows steep mountain ranges such as the Himalaya and Andes, like bare skeletons from which soil is lost as fast as it is formed (plate 5.1). In studies of geological erosion on a continental scale, FOURNIER (1962) quantified macro-relief as shown in figure 5.2.

For assessment of sediment yield from river basins, several parameters have been used such as average stream gradient, change in elevation across and along the basin, and various other geometric measurements.

Accelerated erosion is usually more concerned with micro-relief, i.e. at field scale, and this has been thoroughly investigated. The steepness of the slope is the dominant factor. This affects the downslope movement by splash, and EKERN (1951) suggested the expression: *downslope splash as a % of total splash = 50 + slope (%)*. The erosive power of surface runoff is even more affected by increased slope. This, and the effect of the length of slope, which allows a progressive build-up of both volume and velocity, are discussed in section 7.2.2 in connection with the Universal Soil Loss Equation.

The geometry of variable slope hillsides has also been studied (YOUNG and MUTCHLER 1969). A uniform slope loses more soil than a concave

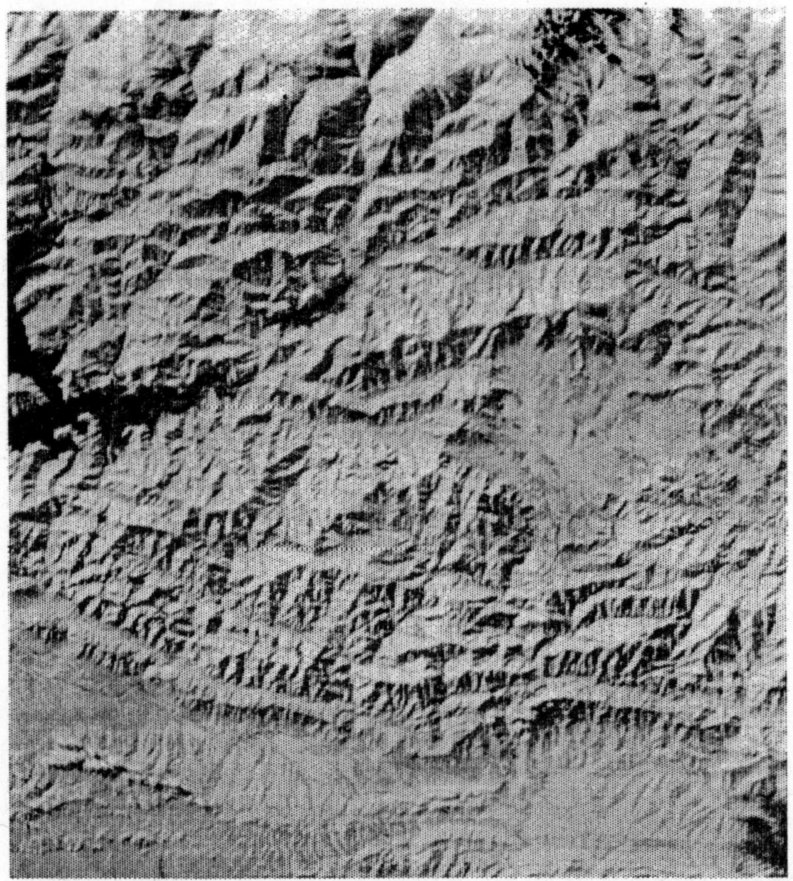

Plate 5.1 Satellite imagery of the Himalaya mountains in Northern India shows them to be rock skeletons where the soil has eroded as fast as it is formed (NASA)

slope but less than a convex slope as shown in figure 5.2. The reason is that the greatest volume and velocity of the runoff which occurs at the bottom end of the slope, operates on the steepest part of the convex slope but on the flattest part of the concave slope.

5.2.4 Land management

The amount of soil erosion which occurs under given conditions is influenced not only by the properties of the soil, but also by the

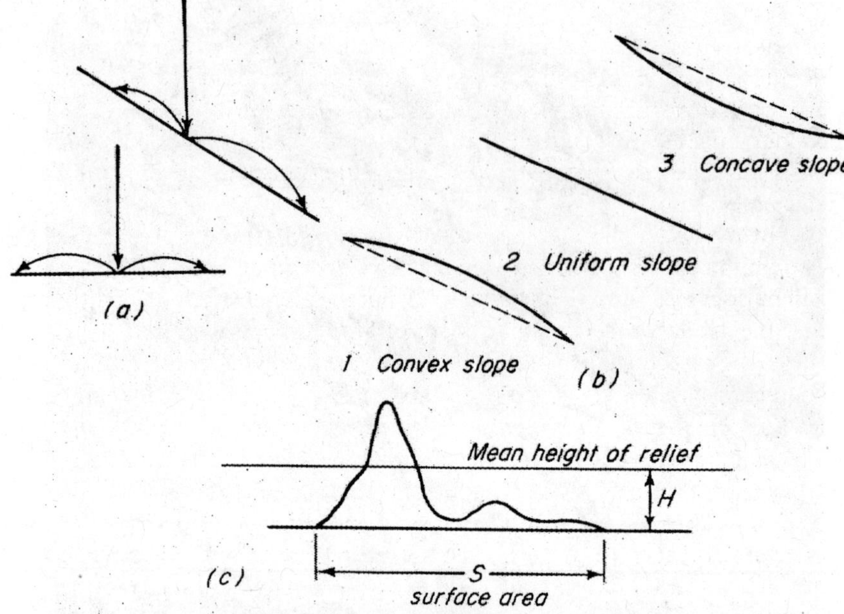

Figure 5.2 The effect of topography on erosion
(*a*) Micro-scale. Net downhill movement by splash increases with slope. EKERN (1951) suggested % splash downhill = 50 + slope %.
(*b*) Field scale. Soil movement on a uniform slope (*2*) is greater than on a concave slope (*3*) but less than on a convex slope (*1*) (YOUNG and MUTCHLER 1969)
(*c*) Macro-scale. The effect of regional topography was quantified as $\frac{H^2}{S}$ by FOURNIER (1962) where H is the average height of the relief and S the surface area

treatment or management it receives. A soil might lose say 400 tons per hectare per year when used for the cultivation of row crops with the rows running straight up and down the steepest slope, while identical soil under well-managed pasture would only lose as many kilograms per hectare. The difference in erosion caused by different management of the same soil is much greater than the difference in erosion from different soils given the same management. In fact erosion is influenced more by management than by any other factor, and management includes both the broad issues of land management and the detailed decisions of crop management. The 'best' land management might be defined as the most intensive and productive use of which the land is capable without causing any degradation. It is perhaps worth stressing the constructive nature of this approach. Conservation policies of land use today must be positive and encouraging, not restrictive. There is no point in preserving soil and not using it – the demand of the world is for its resources to be used as efficiently as possible without waste.

When there is sufficient land available to allow choice between land use options, an effective planning tool is *Land Capability Classification*. This will be discussed further in section 9.1.5, but the essential features are that first the available facts about a piece of land are collected in a special survey. These include items such as the soil type, depth, drainage characteristics, and the slope of the land, which can all be measured or assessed in the field. On the basis of these measured facts the land is then allocated into one of eight classes of land. These classes reflect the risk of erosion and indicate the combination of management practices which will be required for the land to be used productively and sustainably. This system was pioneered by the United States Soil Conservation Service but has been successfully adopted, often with local variations, in other countries.

A problem is that the method assumes that there is enough land to allow choice of use, and this is a luxury not always available in developing countries. In a country with high population pressure, and a high proportion of steep land, it is not practical to restrict cultivation to gently sloping lands. At the farm scale, Capability Classification is an excellent aid to planning the use of a 1 000 hectare farm, but not any help to the peasant farmer whose 1 hectare farm is theoretically unsuitable for cultivation.

5.2.5 Crop management

In the same way that erosion is greatly affected by different kinds of land use, so, within any particular land use, there can be large variations in the amount of erosion depending upon the detailed management of the crops. This will be discussed in more detail in section 11.2, but to give a simple example, the loss of soil from two identical adjacent experimental plots was found to be 15 times greater from the plot with the badly-managed crop of maize than from the plot with a well-managed maize crop (HUDSON 1957). Such differences in erosion due to management are probably most spectacular under row crops, but highly significant differences have also been recorded during experiments with natural grassland, established pasture, forest, woodland, and indeed every form of land use. The old-fashioned concept that crops can be classified as *soil building* or *soil robbing* was seriously challenged years ago as soon as scientific principles were applied to the use of fertilizers. The idea has been refuted by erosion studies in recent years which have shown that the run-down infertility often associated with certain crops is due more to the erosion which they cause than the removal of plant nutrients by the crop. Naturally close-growing crops such as grass will tend to cover and protect the soil, and row crops such as cotton and maize will tend to give less protection, but these general trends can be completely reversed by management. It has been shown by experiment that a well-managed well-grown crop of maize can minimize erosion and build up the soil, and that a badly managed pasture can run down the productivity where there are serious losses of soil and plant nutrients. In fact crop management for erosion control can be summarized in the statement: 'Erosion depends not on what crop is grown, but on how it is grown'. This topic will be expanded in chapter 11.

5.2.6 *Control measures*

This analysis of the factors influencing the erodibility of the soil has led the discussion to the question of how erosion can be controlled. Erosion is a function of erosivity and erodibility, and the erosivity depends entirely on rainfall so is outside our control. Erodibility depends partly on soil properties which again we cannot change, at least only to a small extent, and to a greater extent on land use and crop management, both of which are entirely under man's control.

In section 2.3.2 an equation for soil loss was presented in which $A = RKLSCP$.

In this form R is the erosivity, and all the other factors are components of erodibility. K represents the influence of the physical soil properties and, like the rainfall erosivity R is largely beyond control. The factor S representing slope, and L representing length of slope will only be altered by changing the land surface, for example by bench terracing (which would change S) or by contour terraces (which would change L). The conservation practices factor P is affected by a combination of land management and crop management, so a considerable degree of control is exercised at this point in the equation. The cropping practices factor C depends solely on the crop management, and this is the point where the most effective control measures are brought to bear. The numerical effects of these different aspects of erosion will be discussed in chapter 7.

References

ARMSTRONG, J. L. 1990 Runoff and soil loss from bare fallow plots at Inverell, New South Wales, *Australian Journal of Soil Resources*, 28, 659–675

ADAMS, J. E., D. KIRKHAM and D. R. NIELSEN 1957 A portable rainfall simulator infiltrometer and physical movement of soil in place, *Proceedings of Soil Science Society of America*, 21, 5, 473–477

BARNETT, A. P. and J. S. ROGERS 1966 Soil physical properties related to runoff and erosion from artificial rainfall, *Transactions of the American Society of Agricultural Engineers*, 9, 1, 123–125, 128

BOSAZZA, V. L. 1953 On the erosibility of soils, *African Soils*, 2, 3, and 4, 337

BOUYOUCOS, G. J. 1935 The clay ratio as a criterion of susceptibility of soils to erosion, *Journal of the American Society of Agronomy*, 27, 738–741

BRYAN, R. B. 1968 The development, use and efficiency of indices of soil erodibility, *Geoderma*, 2, 1, 5–26

BRYAN, R. B. and J. DE PLOEY 1983 Comparability of soil erosion measurements with different laboratory simulators, in *Rainfall Simulation, Runoff, and Soil Erosion, Catena Supplement 4*, 33–56

DANGLER, E. W., S. A. EL-SWAIFY and A. P. BARNETT 1975 Erosion losses from Hawaii soils under simulated rainfall, *Research Bulletin 181*, Hawaii Agricultural Experiment Station, University of Hawaii

DANGLER, E. W., S. A. EL-SWAIFY and A. K. F. LO 1987 Predicting the erodibility of tropical soils, in *Soil Conservation and Productivity*, edited by I. Pla Sentis, Soil Science Society of Venezuela

EKERN, P. C. 1951 Raindrop impact as the force initiating soil erosion, *Soil Science Society of America Proceedings*, 15, 7–10

EL-SWAIFY, S. A., E. W. DANGLER and C. L. ARMSTRONG 1982 *Soil Erosion by Water in the Tropics*, University of Hawaii, Research Extension Series 024

ELWELL, H. A. 1986 Determination of erodibility of a subtropical clay soil: a laboratory rainfall simulator experiment, *Journal of Soil Science*, 37, 345–350

FOURNIER, F. 1962 *Map of Erosion Danger in Africa South of the Sahara*, Commission for Technical Cooperation in Africa, EEC

HAMILTON, G. H. 1977 The Assessment of Soil Erodibility, *Journal of the Soil Conservation Service of New South Wales*, 33, 2, 106–107

HENIN, S. 1963 L'appréciation des propriétés physiques du sol, *Annales Gembloux*, 3, 631–633

HUDSON, N. W. 1957 Erosion Control Research – Progress report on experiments at Henderson Research Station 1953–1956, *Rhodesian Agricultural Journal*, 54, 4, 297–323

KHATRI CHETRI, T. B. and D. J. PAINTER 1971 Erodibility of New Zealand Soils, *Journal of Hydrology (NZ)*, 10, 1, 49–58

LAL, R. 1988 Erodibility and Erosivity, chapter 7, in *Soil Erosion Research Methods*, edited by R. Lal, Soil and Water Conservation Society, Ankeny, Iowa

LOCH, R. J. and C. J. ROSEWELL 1992 Laboratory methods for measurement of soil erodibilities (K factor) for the Universal Soil Loss Equation, *Australian Journal of Soil Research*, 30, 233–248

LUTZ, J. E. 1934 The physico-chemical properties of soil affecting soil erosion, *Missouri Agricultural Experiment Station Research Bulletin 212*

MCCALLA, T. M. 1944 Waterdrop method of determining stability of soil structure, *Soil Science*, 58, 117

MEYER, L. D. and D. L. MCCUNE 1958 Rainfall simulator for runoff plots, *Agricultural Engineering*, 39, 10, 644–648

MIDDLETON, H. E., C. S. SLATER and H. G. BYERS 1934 The physical and chemical characteristics of the soil from the erosion experiment stations, *Tec ical Bulletin 430*, United States Department of Agriculture, Washington DC

MOLDENHAUER, W. C. 1965 Procedure for studying soil characteristics by using disturbed samples and simulated rainfall, *Transactions of American Society of Agricultural Engineers*, 8, 1, 74–75

NAMBA, S. 1952 An experiment on Erodibility Index, *Journal of Japanese Forestry Society*, 34, 113–117

OLSON, T. C. and W. H. WISCHMEIER 1963 Soi' erodibility evaluation for soils on the runoff and erosion stations, *Proceedings of Soil Science Society of America*, 27, 5, 590–592

PAEZ, M. L. and I. PLA 1987 Development and testing of erodibility indices for selected Venezuelan soils, in *Soil Conservation and Productivity*, Vol 2, 810–821

PEELE, T. C. 1937 The relation of certain physical characteristics to the erodibility of soils, *Proceedings of the Soil Science Society of America*, 2, 97–100

PEREIRA, H. C. 1956 A rainfall test for structure in tropical soils, *Journal of Soil Science*, 7, 1, 68–74

RAI, K. D., W. A. RANEY and H. B. VANDERFORD 1954 Some physical factors that influence soil erosion and the influence of aggregate size and stability on growth of tomatoes, *Proceedings of the Soil Science Society of America*, 18, 486–9

ROOSE E. J. 1977 Application of the Universal Soil Loss Equation in West Africa, in *Soil Conservation and Management in the Humid Tropics*, edited by D. J. Greenland and R. Lal, Wiley, Chichester, Sussex

ROOSE, E. and J. PIOT 1984 Runoff erosion and soil fertility restoration on the Mossi Plateau (Central Upper Volta), in *Challenges in African Hydrology and Water Resources*, edited by D. W. Walling, S. S. D. Foster and P. Wurzel, IAHS publication 144, 485–498

USDA-ARS 1994 Predicting soil erosion by water: A guide to conservation planning with the Revised Universal Soil Loss Equation (RUSLE), *Agricultural Handbook 703*, National Technical Information Service, Springfield, Virginia

VANELSLANDE A., P. ROSSEAU, R. LAL, D. GABRIELS and B. S. GHUMAN 1984 Testing the applicability of a soil erodibility nomogram for some tropical soils, in *Challenges for African Hydrology and Water Resources*, IAHS publication 144, 463–473

VOZNESENSKY, S. and A. B. ARTSRUUI 1940 A laboratory method for determining the anti-erosion stability of soils, Problems of erosion resistance of soils, *Tiflis*, 18–33, 1940 (in Russian). Abstract, in *Soils and Fertilizers*, 10, 289, 1947

WISCHMEIER, W. H., D. D. SMITH and R. E. UHLAND 1958 Evaluation of the factors in the soil loss equation, *Agricultural Engineering*, 39, 8, 458–462, 474

WISCHMEIER, W. H., C. B. JOHNSON and B. V. CROSS 1971 A soil erodibility nomograph for farmland and construction sites, *Journal of Soil and Water Conservation*, 26, 5, 189–192

YODER, R. E. 1936 A direct method of aggregate analysis of soil and a study of the physical nature of erosion losses, *Journal of the American Society of Agronomy*, 28, 337–351

YOUNG, R. A. and C. K. MUTCHLER 1969 Soil movement on irregular slopes, *Water Resources Research*, 5, 1084–1089

YOUNG, R. A., M. J. M. ROMKENS and D. K. MCCOOL 1990 Temporal variations in soil erodibility, in *Soil Erosion – experiments and models*, edited by R. B. Bryan, *Catena Supplement 17*, 41–53, Cremlingen, Germany

6

The estimation of surface runoff

6.1 Quantities and rates of runoff

Before a start can be made on the design of the channels and ditches and other works which can deal with surface runoff we need to know the probable quantity of water. If the object is to impound or store the runoff then it may be sufficient to know the total volume of water to be expected, but the usual conservation problem is conveying water from one place to another, and in that case the rate of runoff is more important, and particularly the maximum rate at which runoff is likely to occur. That is the flow which a channel must be designed to accommodate.

In a hypothetical catchment area* with an impervious surface and no losses the maximum rate of runoff would be directly proportional to the rate of rainfall. In natural catchments there are other factors: some of the rain is intercepted by vegetation, some infiltrates into the soil, some starts moving over the surface but is trapped in depressions, and some is lost by evaporation. Estimates of rates of surface runoff therefore depend on two processes: estimating the rate of rainfall, and estimating how much of the rainfall becomes runoff.

There are several empirical methods for doing both and for combining the two factors, but because they are empirical two points must be stressed. Firstly, as with all empirical solutions, data, tables, and formulae developed in one situation should not be applied in different conditions without careful checks on whether the method is valid in the new situation. Secondly, each of the three main methods which will be

*There is a variation in terminology. British usage is 'catchment' for the area which catches runoff, and 'watershed' for the boundary of the catchment, i.e. the divide which sheds water on either side. American usage is 'watershed' instead of 'catchment'. Both words are used interchangeably in this chapter according to the usage in the source literature.

Table 6.1 Values of runoff coefficient *C*

Topography and vegetation	Soil texture		
	Open sandy loam	Clay and silt loam	Tight clay
Woodland			
Flat, 0–5 per cent slope	0.10	0.30	0.40
Rolling, 5–10 per cent slope	0.25	0.35	0.50
Hilly, 10–30 per cent slope	0.30	0.50	0.60
Pasture			
Flat	0.10	0.30	0.40
Rolling	0.16	0.36	0.55
Hilly	0.22	0.42	0.60
Cultivated			
Flat	0.30	0.50	0.60
Rolling	0.40	0.60	0.70
Hilly	0.52	0.72	0.82
Urban areas	30% of area impervious	50% of area impervious	70% of area impervious
Flat	0.40	0.55	0.65
Rolling	0.50	0.65	0.80

From SCHWAB, ELLIOTT, FANGMEIER and FREVERT, *Soil and Water Conservation Engineering*, Wiley, New York

described has a number of factors, but the factors are not interchangeable, and each method must be used as a whole.

6.2 Ground conditions in the catchment

The proportion of rain which becomes runoff depends on many factors: the topography, the vegetation, the infiltration rate, the soil storage capacity, and the drainage pattern are just some of them. There are also many ways of quantifying the effects, of which three will be described.

6.2.1 The runoff coefficient

The simplest method is to use a single coefficient which represents the ratio of the rate of runoff to the rate of rainfall. If half of the rainfall is 'lost' by infiltration, etc, and the other half appears as runoff, then the coefficient, denoted by *C* is 0.5. Some values of *C* are shown in table 6.1.

These coefficients are combinations of three topography factors (flat, rolling, or hilly), four categories of land use, and three broad soil descriptions. All the other variables are ignored. The most that can be done to cater for different catchment conditions is to estimate the proportions of the catchments which have different coefficients and then to strike a weighted average. For example, if a catchment were 40% hilly woodland on sandy loam, and 60% rolling cultivated clay loam, then using the figures of table 6.1 the calculation would be

$$\frac{4}{10} \times 0.30 + \frac{6}{10} \times 0.60 = 0.48$$

and this figure would be used as the coefficient for the whole catchment.

The advantage of this method is that it is simple and easy to use. It is used in the *Rational method* discussed in section 6.4.1.

6.2.2 Catchment characteristics

This method is sometimes called *Cook's method*, after the engineer of the US Soil Conservation Service who developed it. It is also known as the ΣW method (pronounced sigma W) because Σ is the mathematical symbol indicating the summation of several values, and the method consists of summing numbers, each of which represents the extent to which runoff from the catchment will be influenced by a particular characteristic.

The letter *W* arises from the American usage of *watershed*, and an African variation on the method is called the *Catchment Characteristics method* (*CC*) after British usage of *catchment*.

The effect of four features is considered in Cook's method: the relief, the soil infiltration, the vegetal cover, and the surface storage. Each of these is considered in turn, and the condition of the watershed is compared with four descriptions shown in table 6.2. The description is chosen which most nearly fits the watershed, and the corresponding number (shown in brackets) is noted. If the catchment condition lies somewhere between two adjacent descriptions in the table, intermediate values may be interpolated. The arithmetic total of the four numbers is the watershed characteristic and will lie between extreme values of 100 (if the highest number has been chosen for each of the four parameters) and 25 (if the lowest value has been chosen in each case).

A modification of this method was developed for African conditions. Surface storage was found to have little effect on peak rates, so a three factor combination was used, assessing the features: cover, soil type and

Table 6.2 Catchment characteristics for Cook's method (USA)

	(100) *Extreme*	(75) *High*	(50) *Normal*	(25) *Low*
Relief	(40) Steep, rugged terrain, with average slopes generally above 30%.	(30) Hilly, with average slopes of 10–30%.	(20) Rolling, with average slopes of 5 to 10%.	(10) Relatively flat land, with average slopes of 0 to 5%.
Soil infiltration	(20) No effective soil cover, either rock or thin soil mantle of negligible infiltration capacity.	(15) Slow to take up water; clay or other soil of low infiltration capacity, such as gumbo.	(10) Normal, deep loam with infiltration about equal to that of typical prairie soil.	(5) High; deep sand or other soil that takes up water readily and rapidly.
Vegetal cover	(20) No effective plant cover; bare or very sparse cover.	(15) Poor to fair; clean-cultivated crops or poor natural cover; less than 10% of drainage area under good cover.	(10) Fair to good; about 50% of drainage area in good grassland, woodland, or equivalent cover; not more than 50% of area in clean-cultivated crops.	(5) Good to excellent; about 90% of drainage area in good grassland, woodland or equivalent cover.
Surface storage	(20) Negligible; surface depressions few and shallow; drainageways steep and small; no ponds or marshes.	(15) Low; well-defined system of small drainageways; no ponds or marshes.	(10) Normal, considerable surface depression storage; lakes, ponds and marshes less than 2% of drainage area.	(5) High; surface depression storage high; drainage system not sharply defined.

From *Engineering Handbook for Farm Planners; Upper Mississippi Valley Region III*, United States Soil Conservation Service, 1953

Table 6.3 Catchment characteristics (African method)

Cover		*Soil type and drainage*		*Slope*	
Heavy grass	10	Deep, well drained soils	10	Very flat to gentle	5
Scrub or medium grass	15	Deep, moderately pervious soil	20	Moderate	10
Cultivated lands	20	Soils of fair permeability and depth	25	Rolling	15
Bare or eroded	25	Shallow soils with impeded drainage	30	Hilly or steep	20
		Medium heavy clays or rocky surfaces	40	Mountainous	25
		Impervious surfaces and waterlogged soils	50		

Select the most appropriate factor from each of these three tables and add them together.
Example Heavy grass (10) on shallow soils with impeded drainage (30) and moderate slope (10): CC = 10 + 30 + 10 = 50.

drainage, and slope. The numerical weightings are shown in table 6.3, and the application of catchment characteristics is shown in section 6.4.2.

Note that the catchment characteristic *CC* is only for use in Cook's method, and is not interchangeable with runoff coefficient *C*, which is for use only in the Rational method.

6.2.3 *Runoff curve numbers*

This is an extension of Cook's method which allows for variations in the physical conditions of a catchment and also the land use. As in Cook's method, four variables are considered, and in each case a selection has to be made from a list of options. Ten categories of land use or cover are offered, as shown in the first column of table 6.4, with a choice of appropriate soil conservation practices such as contouring (i.e. planting on the contour) and terracing (i.e. use of graded channel terraces). The hydrologic condition of the catchment is graded good, fair, or poor, and a subjective assessment of this factor is called for. For arable land, the hydrologic condition reflects whether the rotation will encourage infiltration and promote a good tilth. For grassland it is assessed on the density of the vegetative cover, *eg* more than 75% cover is 'good', while less than 50% is 'poor'. For forest lands the criteria are the depth of litter and humus, and the compactness of the humus. Finally the soil is designated as one of four major hydrologic soil groups described in table 6.5.

A disadvantage of this approach is that it relies on subjective (i.e. non-

Table 6.4 Runoff curve numbers

Land Use or Cover	*Treatment or Practice*	*Hydrologic Condition*	*Hydrologic Soil Group* *A*	*B*	*C*	*D*
Fallow	Straight row	—	77	86	91	94
Row crops	Straight row	Poor	72	81	88	91
	Straight row	Good	67	78	85	89
	Contoured	Poor	70	79	84	88
	Contoured	Good	65	75	82	86
	Terraced	Poor	66	74	80	82
	Terraced	Good	62	71	78	81
Small grain	Straight row	Poor	65	76	84	88
	Straight row	Good	63	75	83	87
	Contoured	Poor	63	74	82	85
	Contoured	Good	61	73	81	84
	Terraced	Poor	61	72	79	82
	Terraced	Good	59	70	78	81
Close-seeded legumes or rotation meadow	Straight row	Poor	66	77	85	89
	Straight row	Good	58	72	81	85
	Contoured	Poor	64	75	83	85
	Contoured	Good	55	69	78	83
	Terraced	Poor	63	73	80	83
	Terraced	Good	51	67	76	80
Pasture or range		Poor	68	79	86	89
		Fair	49	69	79	84
		Good	39	61	74	80
	Contoured	Poor	47	67	81	88
	Contoured	Fair	25	59	75	83
	Contoured	Good	6	35	70	79
Meadow (permanent)		Good	30	58	71	78
Woods (farm wood-lots)		Poor	45	66	77	83
		Fair	36	60	73	79
		Good	25	55	70	77
Farmsteads		—	59	74	82	86
Roads		—	74	84	90	92

From SCHWAB, ELLIOTT, FENGMEIER and FREVERT *Soil and water conservation engineering*, Wiley, New York

Table 6.5 Hydrologic soil groups

Hydrologic Soil Group	*Runoff Potential*	*Infiltration when wet*	*Typical Soils*
A	Low	High	Excessively drained sands and gravels
B	Moderate	Moderate	Medium textures
C	Medium	Slow	Fine texture or soils with a layer impeding downward drainage
D	High	Very slow	Swelling clays, claypan soils or shallow soils over impervious layers

measurable) assessment as well as factual criteria. The variables are combined as in table 6.4 to give the curve number which can range from 25 to 100. Like the runoff coefficient, a weighted average can be calculated when conditions vary in the catchment. The application of runoff curve numbers to estimating peak runoff is shown in section 6.4.3.

6.3 Rainfall

6.3.1 Intensity and duration of rainfall

It is common experience that the most severe rainfall only lasts for a short time. A storm which lasts for several hours will usually give a greater total amount of rain than a storm which lasts a few minutes, but the average rate of rainfall, expressed in mm per hour, will be less than the average rate for the short storm. The length of a storm is called its *duration*, and the relationship between intensity and duration is of the form shown in figure 6.1. The actual shape of the curve, and the numerical values of intensity will vary for different climates, but for all countries the general pattern will be the same, and can be constructed by drawing a trend line through a large number of points plotted from the records of individual storms. The curve can be expressed mathematically in several forms, one of the most common being

$$I = \frac{a}{t+b}$$

where I is the average intensity of the storm in mm/h, t is the duration of the storm in minutes, and a and b are constants.

Figure 6.1 is constructed from values of 6 000 for a, and 50 for b. This relationship is not suitable for short durations of less than 5 minutes when intensities can be higher than the maximum value of 120 mm/h given by this formula.

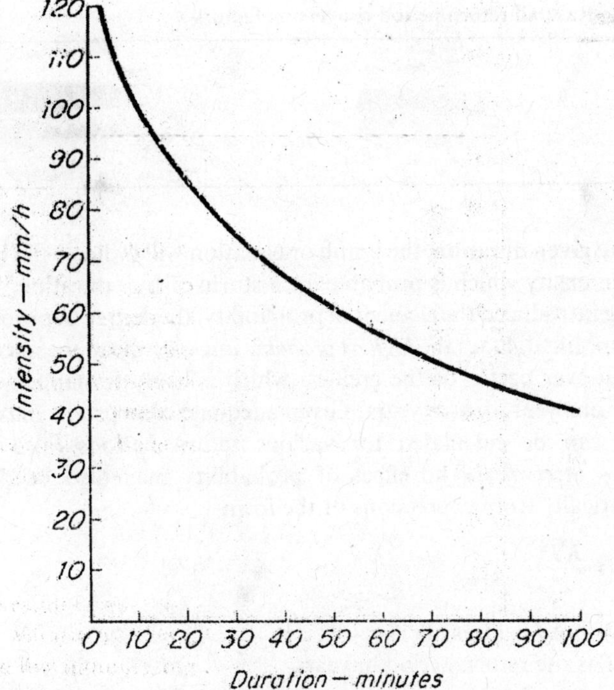

Figure 6.1 The relationship between intensity and duration

Another mathematical solution for the intensity/duration relationship is offered by the INSTITUTE OF HYDROLOGY (1975) in the form

$$I = \frac{Io}{(1+BD)^n}$$

where I is the average intensity in mm/h for 5-year return periods.
Io is the 'instantaneous' intensity or the maximum sustained for very short periods of about 15 seconds. In the UK the values range from 145–170 mm/h.
B is a constant which varies with the mean annual rainfall, and ranges from 15 to 45.
D is duration in hours.
n is the 'continentality' factor, and is low for a rainfall regime with high total but long periods of low intensity rain, and high for more stormy rainfall patterns. The range is 0.44 to 0.78.

Table 6.6 Rainfall return period conversion factors

2 years	0.75
5 years	0.85
10 years	1.00
25 years	1.25
50 years	1.50

For any given duration, the graph or equation will indicate the highest average intensity which is probable for a storm of that duration.

Having introduced the element of probability, the degree of probability must be specified. Does the *highest probable intensity* mean the maximum which can ever occur, or the greatest which is likely to happen over a period of one year, or two years? Given adequate records, the maximum intensity can be calculated for various return periods (also called *recurrence intervals*). The effect of probability may also be derived mathematically from expressions of the form

$$I=\frac{KT^x}{t^n}$$

where I is intensity
T is the return period in years
t is the duration in minutes
K, x and n are all constants.

Any calculations involving rainfall probability must be related to a chosen return period. For engineering structures the return period is linked to the seriousness of the damage which would result in the event of failure – in effect the choice of return period is the choice of the safety factor. So a small earthen farm dam, which could be repaired without much trouble or expense, might be designed to withstand the 10-year flood, but a large storage dam for an irrigation scheme would be designed to cope with a 50-year flood. And if the failure of a major hydroelectric dam would mean damage to property and loss of life, then the design return period might be 500 years or more. Table 6.6 shows that there may be only 25% difference between a 2-year and a 10-year rainfall, so 10 years is frequently used for the design of conservation works. This is a conservative approach and appropriate for structures such as concrete weirs or dams because much of the cost is in the design and the mobilization of the equipment, so the extra cost of a 10-year design compared with a 2-year design is less than the 25% difference in the design flood. It therefore is sensible to buy the extra safety margin cheaply. But

Table 6.7 Conversion of a 2-year, 1-hour storm to other design storms

Duration (hours)	*Frequency in years*					
	2	5	10	25	50	100
1	1.00	1.35	1.65	2.00	2.25	2.50
6	1.65	2.25	2.70	3.30	3.70	4.10
24	2.40	3.25	3.95	4.80	5.40	6.00

FAO 1976

we can modify this approach to risk when designing simple conservation works for the small-scale subsistence farmer. For him the main object may be to minimize the investment input of labour or any other resource. From the farmer's point of view the 'best' design may not be the one which is most cost-effective, but the best he can afford, so he builds a small structure and takes a chance on having to rebuild it if it is washed out in a heavy storm.

6.3.2 *Design storm*

Many hydrological analyses are based on what would happen during a theoretical *design storm* which is a storm of given duration and probability. The 10-year, 1-hour rainfall is the maximum rainfall probable during a 1-hour period with a 10-year return period. When, as frequently happens, the available rainfall data is not in the form required, it is possible to compare the data with more detailed records from another region of similar rainfall, and then to generate conversion factors which will allow the local data to be extrapolated into the required form. An excellent example of this technique is the work of the US SCS in North Africa (FAO 1976). Similarity of the climate on the North African coast and the west coast of the United States of America allowed the construction of the conversion table 6.7 so that a required design storm (in that case the 25-year, 6-hour storm) could be calculated from the available data which was in the form of the 2-year, 1-hour rainfall. This is a powerful tool, but requires caution and experience. The authors emphasise that table 6.7 is an example and should not be used in other conditions.

6.3.3 *Time of concentration*

The storm duration which will correspond with the maximum rate of runoff is known as *the time of concentration* or *the gathering time*. It is defined as the longest time taken for water to travel by overland surface flow from any point in the catchment to the outlet. The reason why this

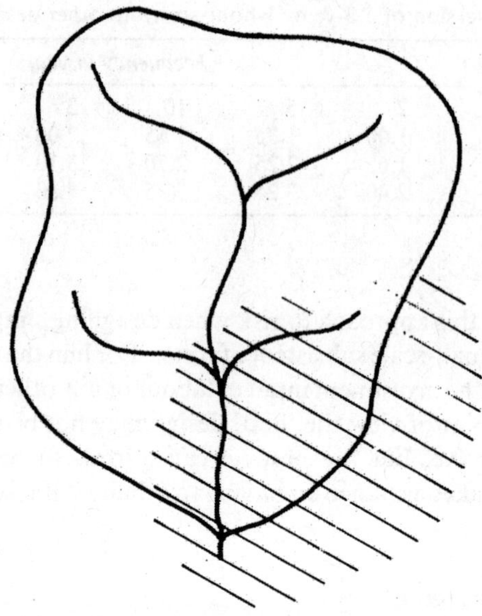

Figure 6.2 A heavy storm on part of the watershed does not give the maximum flow

time corresponds with the maximum flow is best illustrated by considering the catchment shown in figure 6.2. If a severe but localized storm falls in the lower shaded part of the catchment the runoff will be proportional to the product of the intensity and the area on which the storm falls. The intensity could be high but only a portion of the catchment area will be yielding runoff. If a more widespread belt of rain covers the whole area the intensity will be lower, but the whole catchment area will be yielding runoff. Hydrological studies show that for normal catchments this second situation, of a storm covering the whole catchment, always gives a greater maximum rate of runoff. Maximum runoff will therefore result when the whole catchment is yielding runoff. Since the intensity/duration curves show that intensity decreases as duration increases, the maximum rate of rainfall, and hence the maximum rate of runoff, will occur in a storm with the shortest duration which will still allow the whole catchment to contribute runoff. The shortest time for the whole catchment to contribute is the time it will take water to flow from the point in the catchment which is farthest away in time, hence the definition of concentration time. The longest time may not necessarily be that taken by runoff from the farthest point to reach the outlet, for there may be a nearer point which because of

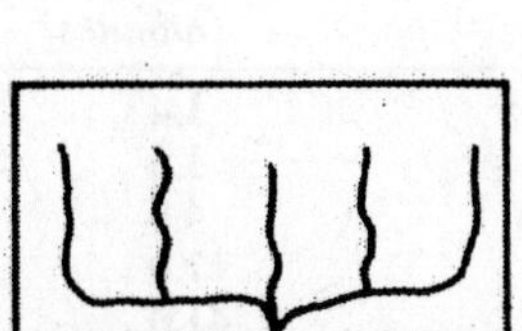

Figure 6.3 A short squat watershed has a shorter gathering time than a long narrow watershed

flatter grades or surface storage has a slower route to the outlet. This possibility is taken care of by the definition specifying the longest time for runoff to flow from a point in the catchment to the outlet.

The main variables affecting the time of concentration of a catchment are:

1 *Size*: the larger the catchment the longer will be the gathering time.

2 *Shape* of the catchment. In figure 6.3 the two catchments have the same area and both have a symmetrical drainage pattern but the longest distance to the outlet is greater in one than in the other. The gathering time will therefore be longer, the corresponding intensity lower, and the maximum rate of runoff less. This is the explanation of the observed fact that, all other factors being equal, long narrow catchments tend to have less flashy floods than square or round catchments.

3 *Topography*: steep topography will cause faster runoff and a shorter gathering time than a flatter catchment.

4 *Ground condition*: soil type, land use, density of cover, and the other factors discussed in section 6.2 affect the rate of runoff in the same way that they affect the amount of runoff, and so are sometimes included in the estimation of time of concentration.

An approximate estimate of the time of concentration can be obtained from area alone, and table 6.8 (from AYRES and COATES 1939) gives some suggested values. Another method which introduces the effect of shape as well as area is the BRANSBY-WILLIAMS formula:

$$T=\frac{L}{1.5D}\sqrt[5]{\frac{M^2}{F}}$$

where T is in hours, L and D in kilometres, M in square kilometres, and F

Table 6.8 Time of concentration from catchment area

Area (acres)	*(hectares)*	*Time of Concentration or Gathering Time—(minutes)*
1	0.4	1.4
5	2.0	3.5
10	4.0	4.0
100	40.5	17.0
500	202.5	41.0
1000	405.0	75.0

(AYRES and COATES 1939)

Table 6.9 Time of concentration for small catchments from Kirpich formula

Average gradient in catchment (per cent)	0.05	0.1	0.5	1.0	2.0	5.0	10.0
Maximum length of flow (m)	*Time of concentration (minutes)*						
100	12	9	5	4	3	2	1.5
200	20	16	8	6	5	4	3.3
500	44	34	17	13	10	8	6.6
1000	75	58	30	23	18	13	10.0
2000	130	100	50	40	31	22	15
3000	175	134	67	55	42	30	22
4000	216	165	92	70	54	38	30
5000	250	195	105	82	65	45	35

in metres per 100 metres. Another formula, popular in America, is that of KIRPICH (1940)

$$T(\text{minutes}) = 0.02L^{0.77}\ S^{-0.385}$$

where

T is the time of concentration in minutes
L is the maximum length of flow in metres
S is the average stream gradient in metres per metre.

Table 6.9 shows some solutions of the Kirpich equation.

When information is available to calculate the runoff curve numbers as described in section 6.2.3 and table 6.4, this can be combined with area and average land slope to estimate time of concentration from figure 6.4.

6.4 Estimating peak rate of runoff

The estimate of probable rainfall discussed in section 6.3 can now be combined with the effect of ground conditions, discussed in section 6.2, to estimate probable maximum rate of runoff. It must be emphasised again that the three optional methods now described are all empirical and must be used as a whole. It is not permissible to combine the ground conditions part of one method with the rainfall estimate of another.

6.4.1 *The rational method*

The popularity of this method was enhanced by a fortunate coincidence of numbers in the old imperial units of feet, inches, and acres. The mathematics of this are that for rain falling at one inch per hour on one acre:

$$43\ 560 \text{ (acre to sq ft)} \times \frac{1}{12} \text{ (in. per hr to ft)} \times \frac{1}{3600} \text{ (hr to secs)} = 1.008 \text{ cu ft per sec,}$$

which for all practical purposes can be taken as unity. The proportion of rain which becomes runoff can then be expressed as a dimensionless ratio, in fact the runoff coefficient C discussed in section 6.2.1, and the equation $Q = CIA$ where

Q is runoff in cubic feet per second
C is the dimensionless runoff coefficient
I is intensity in inches per hour
A is area in acres.

The Rational formula is equally applicable in metric units but an additional constant is required to reconcile the units of time.

$$Q\text{(cubic metres per sec)} = \frac{CIA}{360}$$

where

C is the dimensionless runoff coefficient
I is intensity in mm per hour
A is area in hectares.

The application of the formula consists of selecting appropriate values of C, I, and A. The area A can be measured by survey or from maps or aerial photographs. The value of intensity I is the maximum rate of rainfall, determined as shown in section 6.3.1 from consideration of the time of concentration for the catchment, and the probability.

The main advantage of the Rational Formula is that it can always be used to give an estimate of maximum runoff rates no matter how little recorded information is available. Naturally the precision of the estimate depends upon the precision of the information put in, but even in an undeveloped country with no rainfall or hydrological records at all, the method can be used to give an estimate which will be very much better than the only other alternative which is pure guesswork. The area can always be measured; there are several formulae for time of concentration, and these can be applied to the general formula for intensity if there is nothing better from local records; similarly if there are no measurements of local values of the runoff coefficient, then the published data of other countries will serve. As information on rainfall and runoff is obtained it can be fed in to the formula to give progressively more accurate estimates.

In countries where hydrological experiments have provided more detailed information, more sophisticated methods have been developed for estimating rates of runoff, and these make better allowance for variations in the catchment conditions than is possible in the rational formula.

6.4.2 Cook's method

This method is very simple to apply, but depends on the existence of data recorded from experimental watersheds. An estimate is made of the catchment characteristics as in section 6.2.2, and of the area of the catchment, then from these two variables the runoff is read off directly from a table such as the example shown in table 6.10. This example is based on a 10-year probability for a tropical rainfall with severe storms of high intensity.

Cook's method differs from the rational formula in that it requires some measured values to establish the scales. But this does not mean that its use is restricted to developed countries with long established programmes of experimental hydrology. All that is required as a starting point is a few carefully chosen experiments to establish reference points which will make it possible to make use of the published results from other countries. Table 6.10 was derived in this manner for use in Central Africa from data published by the United States Agricultural Research Service.

Two more variables can be introduced into Cook's method if required. If regional variations in rainfall exist, and the effect on peak runoff can be estimated, this can be allowed for, by applying a conversion factor. For example a geographical region, known to experience less intense rainfall, could have a reducing ratio, say 0.75, to be applied to the figure obtained from table 6.10.

The other optional variable is in the shape of the catchment. We saw in section 6.3.3 that shape affects the time of concentration, and hence the

Table 6.10 Runoff from small catchments

A \ CC	25	30	35	40	45	50	55	60	65	70	75	80
5	0.2	0.3	0.4	0.5	0.7	0.9	1.1	1.3	1.5	1.7	1.9	2.1
10	0.3	0.5	0.7	0.9	1.1	1.4	1.7	2.0	2.4	2.8	3.2	3.7
15	0.5	0.8	1.1	1.4	1.7	2.0	2.4	2.9	3.4	4.0	4.6	5.2
20	0.6	1.0	1.4	1.8	2.2	2.7	3.2	3.8	4.4	5.1	5.8	6.5
30	0.8	1.3	1.8	2.3	2.9	3.6	4.4	5.3	6.3	7.3	8.4	9.5
40	1.1	1.5	2.1	2.8	3.5	4.5	5.5	6.6	7.8	9.1	10.5	12.3
50	1.2	1.8	2.5	3.5	4.6	5.8	7.1	8.5	10.0	11.6	13.3	15.1
75	1.6	2.4	3.6	4.9	6.3	8.0	9.9	11.9	14.0	16.4	18.9	21.7
100	1.8	3.2	4.7	6.4	8.3	10.4	12.7	15.4	18.2	21.2	24.5	28.0
150	2.1	4.1	6.3	8.8	11.6	14.7	18.2	21.8	25.6	29.9	35.0	40.6
200	2.8	5.5	8.4	11.7	15.3	19.1	23.3	28.0	33.1	38.5	45.0	52.5
250	3.5	6.5	9.7	13.2	17.2	21.7	27.0	32.9	39.6	46.9	55.0	63.7
300	4.2	7.0	10.5	14.7	19.6	25.2	31.5	38.5	46.2	54.6	63.7	73.5
350	4.9	8.4	12.6	17.2	23.2	30.2	37.8	46.3	53.8	62.5	71.5	81.0
400	5.6	10.0	14.4	19.4	25.6	33.6	42.2	51.0	60.0	69.3	79.5	90.0
450	6.3	10.5	15.5	21.5	28.5	36.5	45.5	55.5	65.5	76.0	86.5	97.5
500	7.0	11.0	17.0	23.5	31.0	40.5	51.0	62.0	73.0	84.0	95.0	106.5

A is the area of the catchment in hectares, *CC* is the catchment characteristics from table 6.3, and the runoff (in cubic metres per second) is for a 10-year frequency

design intensity, and hence the peak runoff. A simple approach is to assume that the basic data in table 6.10 are for a catchment that is roughly square or round, and to multiply by 0.8 for a long and narrow catchment, or by 1.25 for a broad and short catchment. Any calculation involving rainfall probability must be related to a return period, and the period usually chosen for the design of major conservation works is 10 years. The data are therefore presented for a 10-year return period, and may be converted to other recurrence intervals by factors shown in table 6.6.

6.4.3 *The United States Soil Conservation Service Procedure*

The vast experience of the US-SCS in design procedures has resulted in the development of excellent procedures and design manuals. As with all empirical methods, it requires sensible caution to apply them to other situations, but it can be done, as in the North African example quoted in section 6.3.2 (FAO 1976). The step by step procedure follows, with sample figures in square brackets.

Table 6.11 Unit peak discharge

Time of Concentration	*Peak Discharge*	*Time of Concentration*	*Peak Discharge*	*Time of Concentration*	*Peak Discharge*
(*Hours*)	(m^3/s)	(*Hours*)	(m^3/s)	(*Hours*)	(m^3/s)
0.1 and less	0.337	1.0	0.158	8	0.039
0.2	0.300	1.5	0.120	10	0.034
0.3	0.271	2.0	0.100	12	0.030
0.4	0.246	2.5	0.086	14	0.027
0.5	0.226	3.0	0.076	16	0.025
0.6	0.208	4.0	0.063	18	0.023
0.7	0.195	5.0	0.054	20	0.021
0.8	0.180	6.0	0.048	22	0.020
0.9	0.168	7.0	0.043	24	0.019

* Discharge rate in cubic metres per second from a runoff depth of one millimetre and a drainage area of one square kilometre. (FAO 1976)

1 Determine a design storm from existing rainfall data
[The 2-year 1-hour rainfall is 20.3 mm]
2 Convert to the required design storm using table 6.7
[The peak flow required is the 6-hour storm with 25-year probability. Table 6.7 gives a conversion ratio of 3.3. The 25-year design storm is therefore $20.3 \times 3.3 = 67$ mm]
3 Determine the runoff curve number as in section 6.2.3
[Assume the weighted average curve number (CN) is 86]
4 From figure 6.5 determine the runoff to be expected from the design storm
[For rainfall of 67 mm and CN 86, the depth of the runoff is 35.6 mm]
5 From figure 6.4 determine the time of concentration t_c
[For CN = 86 steep and area 32 ha, $t_c = 0.15$h]
6 From table 6.11 determine the 'unit peak discharge' which is the peak discharge expected from this design storm, expressed in cubic metres per second, for each millimetre of runoff, from a drainage area of one square kilometre
[For $t_c = 0.15$, unit discharge is 0.319 m^3/s]
7 Convert the unit discharge to actual for the given area and runoff depth
[Peak discharge $= 0.319 \times 0.32 \times 35.6 = 3.64$ m^3/s]

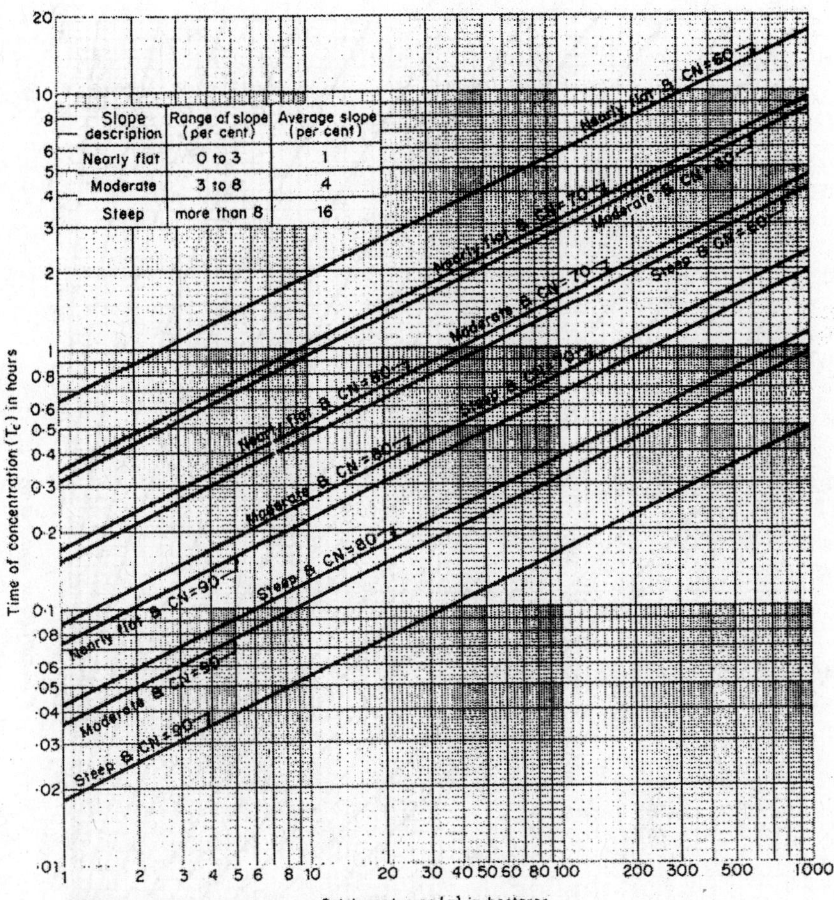

Figure 6.4 Time of concentration for the runoff curve method (FAO 1976)

It must be stressed that this is only an example of how the method can be applied. Tables 6.7 and 6.11, and figures 6.4 and 6.5 are all taken from a particular case on the coast of North Africa (FAO 1976). Application of the method to other areas should only be made after consultation with the Engineering Division, Hydrology Branch, US-SCS.

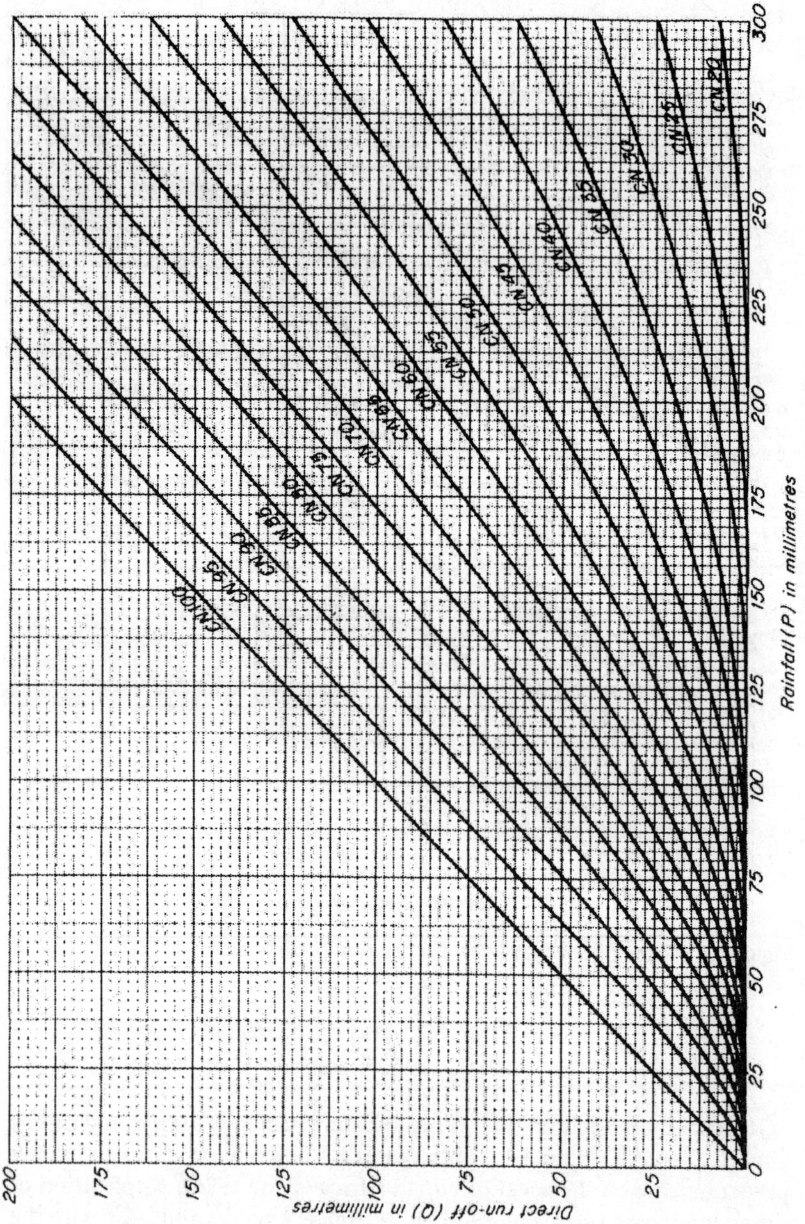

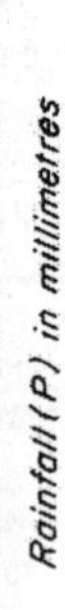

Figure 6.5 Runoff from the design storm

References

AYRES Q.C. and D. COATES 1939 *Land Drainage and Reclamation*, McGraw-Hill, New York

FAO 1976 Hydrology for soil and water conservation in the coastal regions of north Africa in hydrological techniques for upstream conservation, USDA-SCS North Africa project, in *Conservation Guide 2*, edited by S. H. Kunkle and J. L. Thames, FAO, Rome

INSTITUTE OF HYDROLOGY 1975 Flood Studies Report, *Meteorological Studies*, Vol II, National Environmental Research Council

KIRPICH, P. Z. 1940 Time of concentration of small agricultural watersheds, *Civil Engineering*, 10, 362

SCHWAB G., W. ELLIOTT, D. FANGMEIER and R. K. FREVERT 1993 *Soil and Water Conservation Engineering*, 4th edition, Wiley, New York

7

Estimating soil loss

7.1 The purpose and type of models

Sometimes the simplest way to estimate the effect of a physical process is to use an established equation, or formula, or charts and diagrams based on existing knowledge. The current fashion is to call such aids and solutions *models*, and they tend to become increasingly complicated and consequently dependent on computers to handle the mathematical processes.

It is important to identify the exact objective and purpose of models which are designed to estimate soil erosion. Two main divisions are models for prediction and models to assist with explanation. If the purpose is to predict amounts of erosion under alternative management practices then empirical models such as the USLE are efficient and effective. For an understanding of the processes of erosion it is necessary to turn to physically-based or deterministic models. The two objectives should not be cross-threaded onto the wrong kind of model, i.e. the USLE should not be seen as a vehicle for research studies on the mechanics of erosion. Neither should sophisticated simulation models be used as the basis for evaluating conservation practices which are better designed on the basis of experience, and evaluated by direct testing in the field.

The different kinds of models need a short explanation:

An *empirical solution* or model is one based on observation or experiment, and not derived from theory. It fits observed facts, and allows us to predict what will happen in certain circumstances, because we know what has happened before in those circumstances. The reliability of such methods depends on the database of experience: we might say 'This will certainly happen' (because it always does), or 'Will probably happen' (because it usually does), or 'it may happen' (because it sometimes does).

An empirical solution may be a simple approximate relationship or a complex multiple regression equation.

Comparing the estimate obtained from one empirical method with that obtained from another empirical method without new experimental data is totally futile and unscientific, for example comparing EI with other empirical estimates of erosivity. The foundation and starting point of all empirical studies must be physical measurements.

A *factor model* is an empirical model where each of the variables is represented by a quantified factor, and then the factors are combined, for example by adding them up or by multiplying them. Examples are Cook's method for estimating runoff discussed in section 6.2.1, and the Universal Soil Loss Equation for estimating soil loss, discussed in section 7.2.2.

The term *'black-box model'* describes a model where the user feeds in defined items of information and takes out the answer without needing to know or understand what happens inside the black box. Such models are also called *mystery models*. These models can only be operated in the designed direction, that is all the inputs go into one side of the equation, and the answer is the output on the other side of the equation. A regrettably common misuse of one very important black-box model, the Universal Soil Loss Equation, is to try to drive it in reverse and solve the equation for one of the inputs.

Empirical models are created from a database of experience obtained under certain conditions, so the model can be expected to work in those conditions. But it will not necessarily work under different conditions. No model can ever be truly universal – even process-based models need site-specific correction parameters. Ideally, each country or region should design its own prediction models to match its own conditions and to suit its database. Some examples are the Soil Loss Estimation Model for Southern Africa (SLEMSA), designed in Zimbabwe, and the European Soil Erosion Model (EUROSEM).

Procedures derived in the USA are built on a huge database from research, and there is a great temptation to apply them in less developed countries which cannot afford the time and expense to accumulate enough data to create local variations. A certain amount of extrapolation can be useful, for instance some components of the USLE may be borrowed, but this should be done with caution. The possibilities and the pitfalls of transferring USA technology on erosion prediction has been carefully studied (FOSTER *et al* 1982).

A *process-based model* (also called an *analytical component model*), is able to explain mathematically each of the separate physical processes and then combine the separate effects. Because there are so many variables and so many mathematical computations, such models can only be operated through computers.

Stochastic models are useful particularly in hydrological studies and are based on the probabilities of the occurrence of events in a long time series. Both the last two types of model are outside the scope of our present discussion.

The philosophy of different modelling approaches, i.e. analytical component models, stochastic models, and 'mystery black-box' models are reviewed by RENARD (1977), and process-based modelling is reviewed by FOSTER (1990). The progressive development of modelling in the USA is reviewed and explained by LANE *et al* (1992).

7.2 Empirical or black-box models

7.2.1 The development of the Universal Soil Loss Equation

The equation in its present form was not an instantaneous creation, but rather evolved and developed as information became available through research. The starting point for numerical expressions of erosion was the work of ZINGG (1940) when the effects of length of slope and steepness of slope were evaluated. Soon after this, SMITH (1941) defined the concept of a permissible soil loss, and made a first evaluation of a crop factor, and a factor allowing for different degrees of mechanical protection. BROWNING and co-workers in Iowa studied particularly the erodibility of soils, and the effect on erosion of rotations and crop management (1947). About the same time a committee was appointed to integrate previous studies with information then coming forward about the importance of splash erosion. New values were used for several of the variables, and for the first time provision was made for variations in rainfall. The results were sometimes known as the *Musgrave Equation*, after the chairman of the committee, and also known as the *Slope-Practice Equation*, because the slope and farming practice were two of the more important variables (MUSGRAVE 1947). The form of the equation was: E (erosion) = T (soil type) × S (slope) × L (length) × P (agronomic practice) × M (mechanical protection) × R (rainfall). After serving the SCS well for nearly ten years this was replaced in the late fifties by the Universal Soil Loss Equation which, like Wischmeier's Index of Erosivity, was a result of the Purdue study of all the available data from field experiment stations (WISCHMEIER *et al* 1958). Several refinements were added, particularly:

- The new erosivity index allowed more accurately for variations in rainfall from one storm to another, or between seasons.

- It also made it possible to identify local climate, and to allow for crop management techniques related to these.
- The comparison of results from experiment stations on different soils extended the range of erodibility values, and these were redrawn on a new scale.
- Allowance was made for inter-relationships between some of the component parts of crop management such as the level of productivity, crop sequence in the rotation, and the management of residues.

The USLE is presently being replaced in the USA by the Revised Universal Soil Loss Equation (RUSLE) discussed in section 7.2.3, but the simplicity and ease of operation of USLE will ensure its continued use in some countries as a guide to land use planning.

7.2.2 *Using the Universal Soil Loss Equation (USLE)*

The USLE has been a valuable and successful tool for nearly 40 years. Its appeal has led to attempts to use it for purposes it was not designed for, and this has resulted in it sometimes being unjustly criticized. At one time this led the principal author of the system to explain how it should be and should not be used (WISCHMEIER 1976). The purpose is very simple and specific. It provides an estimate of the long-term average annual soil loss from segments of arable land under various cropping conditions. The application of this estimate is to enable farmers and soil conservation advisors to select combinations of land use, cropping practice, and soil conservation practices, which will keep the soil loss down to an acceptable level. In today's terminology we would say to ensure that the farming system is sustainable. Because it was designed for field use it had to be: 'easy to solve and to include only factors whose value at a particular site can be determined from available data. Some potential details and refinements were sacrificed in the interests of utility' (WISCHMEIER 1976).

Some of the things the USLE is not intended to do (and should therefore not be criticized for not doing) are:

- Predicting sediment yield from a watershed, because it does not include deposition and delivery ratios.
- Predicting soil loss from a single storm, because the factors are all long-term averages which smooth out the large variations.
- Predicting soil loss outside the range of its own database (for example the slope factor has only been experimentally deter-

mined up to 16%, and extrapolation beyond this should be tested by experimental studies).

- Predicting soil loss in concentrated channel flow such as large rills and ephemeral gullies.
- Predicting gully erosion or streambank erosion.
- Predicting sediment movement in streams and rivers.
- Predicting the deposition of eroded soil.
- Separating the factors as if they were each independent. As Wischmeier says: 'The relation of a particular parameter to soil loss is often appreciably influenced by the levels at which other parameters are present. To the extent that these interaction effects could be evaluated from existing data, they are reflected in the equation through the established procedures for computing local factor values. Factor *R* reflects the interaction of storm size and rain intensities.' Wischmeier gives some other examples where interaction is partly accounted for, but the basic assumption is that each factor is an independent variable.
- Being used as a precise research tool to study the processes of erosion by treating it as a mathmatical equation which can be solved for one of the input factors, for example by measuring soil loss, estimating all the factors but *K*, and then solving for *K*.

The equation is presented in the form

$$A = R \times K \times LS \times C \times P$$

where

A is the average annual soil loss in tonnes per hectare
R is a measure of the erosive forces of rainfall and runoff
K is the soil erodibility factor – a number which reflects the susceptibility of a soil type to erosion, that is it is the reciprocal of soil resistence to erosion
L is the length factor, a ratio which compares the soil loss with that from a field of specified length of 22.6 metres
S is the slope factor, a ratio which compares the soil loss with that from a field of specified slope of 9%
C is a crop management factor – a ratio which compares the soil loss with that from a field under a standard treatment of cultivated bare fallow
P is the conservation practice factor – a ratio which compares the soil loss with that from a field with no conservation practice, i.e. ploughing up and down the slope.

Converting the USLE for use in metric or SI units is a minefield best avoided if possible. All research data and the analysis which led to the black-box USLE were in foot/pound/acre units, and poking about inside the black box to change the algorithms into a new set of units is likely to cause confusion and error. The first translation was into metric units in *Agricultural Handbook 537* (WISCHMEIER and SMITH 1978), with a supplement for SI units in January 1981, and a detailed proposal for a full-scale conversion to SI by FOSTER *et al* in 1981a. However, the adoption of metrication in the USA has been so hesitant that there is no indication of when it will be generally adopted by farmers and government employees, so RUSLE has been constructed using the old units with a short list of the conversion factors to SI.

In countries accustomed to metric or SI units, one approach would be to use the Australian Soil Loss programme, a national enhancement of the SOILOSS programme which is based on the USLE but implementing some features now included in RUSLE and modifications for local conditions (ROSEWELL and EDWARDS 1988).

7.2.3 *USLE factors*

The factors L,S,C, and P, are each dimensionless ratios which allow comparison of the site being estimated with the standard conditions listed above.

R, the erosivity factor, is computed by the EI method which is the summation for each rainstorm of the kinetic energy (expressed in MJ/ha when using metric units in the USA, or in J/m^2 in Europe), multiplied by the greatest amount of rain in any 30 minute period expressed in cm/h in USA or in mm/h in Europe, as discussed in section 4.3.

K, the erodibility factor, is the average soil loss in tonnes per hectare for each unit of the metric *R* as calculated by the EI method. In effect, the units of *K* are arbitrarily chosen so that when multiplied by *R* in its unconventional units the product is in tonnes per hectare. Values have been tabulated for most American soils, and they may be estimated from the nomograph, figure 5.1.

Although *R* and *C* are average annual factors, there is provision within the system for interaction between them. For example, a period in spring when the soil is newly ploughed and without any plant cover will be dangerous if the spring rains are highly erosive, but not a problem if the spring rains are gentle. Similarly, the effect on erosion of the state of the ground after harvest in the autumn will depend on whether or not the autumn rains are severe. To allow for this possibility the growing season may be divided into periods and *C* and *R* values calculated for each period. The annual combined effect is the sum of the product for each period. Mathematically the effect is the same:

$$C \times R = c_1 r_1 + c_2 r_2 + c_3 r_3 \ldots$$

Extending the USLE to other countries

In the USA the wealth of data allows this calculation to be carried out with precision, and detailed charts and tables are provided in *Agricultural Handbook 537*. However the crop rotations and cropping practices and rainfall distribution pattern are all specific to the USA, and in most other countries it will be more appropriate to use annual values of C and R.

The principle of separately quantifying the rainfall and soil erodibility factors in a format of multiplying them together is equally valid anywhere. There is no theoretical requirement for this format and one could construct a model in which the effects were combined by addition instead of multiplication, but this format is simple and it works. Similarly the concept of dimensionless ratios to compare a particular situation with a standard set of conditions is also transferable. It would be possible to leave out any of the present ratio factors or to insert others but this would require a major research programme and the more practical approach is to try to establish local values of the existing factors.

It would be impractical for other countries to attempt to acquire a database comparable with that of the USA which now comprises something like 10 000 plot years of experimental records. It is therefore worth considering which of the principles and which of the USA-derived factors can be transferred for use in other countries.

The erosivity factor R is empirical but the concept of basing it on rainfall energy and intensity has been confirmed in many countries. When attempts were first made to apply the USLE in the tropics alarmingly high predictions of soil loss were calculated because the database of rainfall energy values obtained in the USA did not include the high intensities of tropical rainfall, and extrapolation induced major errors, discussed in section 4.3.1. Later studies of high intensity rainfall in countries around the world established more realistic values for the energy of high intensity rainfall and these have now been incorporated into the USLE. In its present form, calculation of erosivity values by the EI method are likely to be reasonable for most rainfall regimes.

There is still some slight anxiety about applying the USLE under tropical and sub-tropical rainfall regimes which include high-intensity convective thunderstorm rainfall. The USLE estimates the long-term average annual soil loss, assuming that over and under estimates of soil loss in individual storms will balance out over a long period (WISCHMEIER 1976). But it was Wischmeier's view (1962) that the overall pattern in America is for the greater part of the total erosion damage to be done by frequent small storms. Later studies, discussed in section 4.5.3, have shown that in tropical and sub-tropical rainfall it is the infrequent torrential storms which do most of the damage (LITTLE in Missouri 1940, GREER in Mississippi 1971, BARNETT *et al* in Georgia 1958, HUDSON in Zimbabwe 1957). Averaging the effect of separate storms may be less effective when most of the total comes from extreme events.

The soil erodibility factor K is least likely to be transferable as several studies have shown that the USLE nomograph is not applicable to many tropical and sub-tropical soils (VANELSLANDE *et al* 1984). The four factors fed into the nomograph are the percentage of silt plus very fine sand, the organic matter content, the soil structure, and the permeability. It is probable that the source of the discrepancy is that the silt and very fine sand content and the organic matter content are lower in tropical soils than the medium textured soils of mid-west USA. The only reliable way to establish local values for *K* is to use runoff plots under the standard conditions of bare fallow. It is commonly assumed that once the *K* value has been established for a soil this can be regarded as permanent. This is a useful simplification when the USLE is being used for its proper purpose, but in fact the value can change as a result of soil management; for example, the soil structure can change as a result of cultivation, and the organic matter content may be reduced by cropping or increased by manuring. There is also some evidence of seasonal variations in *K* values, particularly in climates with pronounced wet and dry seasons.

The factors of slope length L and slope steepness S are combined together as shown in figure 7.1 but this is only for convenience and the two factors derive from two separate and different relationships. The USA database for steepness factor *S* extends to 18° but it is quite likely that the physics of fluid flow and sediment transport will not be the same on very steep slopes, and this is a possible area for local investigation. The length factor *L* is less likely to vary and the need to validate this relationship is of fairly low priority. The cover-management factor *C* certainly needs to be researched. The crop rotations and variations of the USA mid-West have been minutely researched and documented and it is clear that the primary purpose of the *C* factor is to reflect how much protection is given to the soil by the vegetative cover. This principle will be the same under any cropping practice, but the tillage and cultivation practices may be quite different from those of the mid-West so local investigation of *C* is desirable.

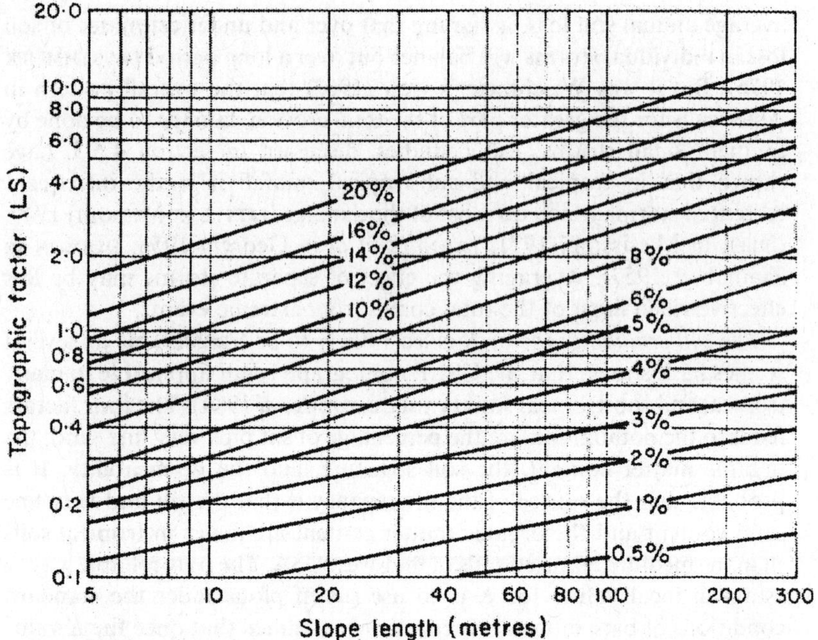

Figure 7.1 The combined Slope-Length factor (WISCHMEIER and SMITH 1978). RUSLE introduces other factors (section 7.2.3)

The conservation practice P is rather crude in the USLE compared with the precision with which the other factors are calculated. One of the reasons is that the effect of major surface manipulation such as graded channel terraces cannot be satisfactorily evaluated with small plots as discussed in section 8.3. FOSTER *et al* (1982) suggest that most of the mechanical practices such as contouring, strip cropping, terraces, and contour furrows, which are used to support protection provided by crop rotation, canopy cover, and residue mulches, are probably transferable. There are no values established in the USA for trash lines, grass strips, or agroforestry practices, but these are the subject of research in several countries. Also there are no known values of *P* for bench terracing but it can be argued that the mechanics of erosion on bench terraced land are so different that in its present form the USLE is not appropriate to this situation.

7.2.4 *Revised Universal Soil Loss Equation (RUSLE)*

The USLE is without doubt the most successful and most used prediction model. Successful, because it is extremely efficient for its purpose of assisting farm planning, and also it has the huge advantage of being simple to understand, to explain, and to use. It needs only 'back of an envelope' calculations, or at most a purpose-built slide rule, illustrated in earlier editions of this book, page 202.

When, after 30 years use, proposals began to emerge for its replacement, there were many in favour of the philosophy 'if it ain't broke, don't fix it'. However, there were good reasons for revising USLE, particularly that it could not accommodate new ideas, new results, and new practices that were not available to its original designers. For example:

- Much additional rainfall data has been acquired during the last 30 years.
- New farming practices have emerged such as ridge-tillage systems, and some long-established practices are not accounted for, such as stripcropping.
- USLE cannot accommodate new research results such as the effect of cropland furrow gradients (MEYER and HARMON 1985).
- The equation cannot accommodate practices which significantly reduce the quantity of runoff, such as level channel terraces.
- The equation cannot cope with erosion caused by local concentrated flows (now called *ephemeral gullies*).
- Inter-rill and rill erosion are lumped together, but the slope factor exponent can vary according to the proportion of rill and inter-rill. Neither could USLE cope with rilling down the side of ridges, now known to generate C factors up to 1.5.
- The equation could not readily be extended to other situations, such as erosion on rangeland, or predicting soil loss over shorter time intervals.

In the mid-eighties, arguments for new models won the day, and a two-part programme was adopted by USDA. In the short term a revised USLE would be constructed, suitable for operation on a personal computer, and incorporating as much as possible of the new information but still essentially an empirical model. In parallel with the RUSLE project would be the development of a new generation of technology for prediciting water erosion by a team in the Water Erosion Prediction Project (WEPP) working with other agencies and academic institutions. The goal of WEPP is a process-oriented model, or family of models, that are conceptually superior to the lumped-model RUSLE and are more versatile as to the conditions that can be evaluated. It is expected that at some time in the future the WEPP technology will replace RUSLE (RENARD *et al* 1991).

The main changes in RUSLE (USDA-ARS 1994)

Rainfall erosivity term R

In 1965 isoerodent maps in *Agricultural Handbook 282* only covered the area of the United States of America which is east of the Rocky Mountains, because there were insufficient data available from the western States. The 1978 *Handbook 537* extended the map to the Pacific coast by including estimates based on the relationship between erosivity and the two-year, six-hour rainfall amount as discussed in section 4.3.1. Since then data from more than one thousand locations in the western States have been analyzed, and new values calculated. The isoerodent maps of RUSLE include the western States, with much more detail. Separate maps of the Pacific coastal states show much local variation with pockets of high-erosivity values which were not previously recorded.

Another change is that *R* values are reduced where flat slopes occur in regions of long intense rainstorms because surface ponded water reduces the erosivity of the raindrop impact.

Erodibility factor K

The nomograph for determining erodibility has been retained, but studies have shown that it does not apply to all soils, so erodibility data from around the world have been reviewed, and an equation has been developed that gives a useful estimate of *K* as a function of an average diameter of the soil particles. There is also an alternative algorithm for volcanic soils based on data from Hawaii.

A significant innovation is the incorporation of research which showed that erodibility varies seasonally, and this can be correlated with changes in soil moisture and soil temperature. This is addressed in RUSLE by calculating for 15-day periods the product of appropriate values of *R* and *K* in the same way as was done in USLE for changes in crop cover during the growing season (section 7.2.2).

Another change in *K* values is allowance for rock fragments in the soil which can affect permeability.

Topographic factors L and S

The algorithms for effect of slope steepness (*S*) and length of slope (*L*) have been revised, and the main effects are that computed soil loss on steep slopes is reduced, and also the algorithms account for differences in the ratio of rill to inter-rill erosion. The basic topographic factor (*LS*) applies to uniform slopes, but non-uniform slopes, such as convex, concave, or complex can be accommodated by dividing the slope into segments (usually up to five) and calculating an average *LS* value.

An alternative length-slope factor has been proposed, but not so far incorporated – a dimensionless sediment transport capacity index which is claimed to be 'simpler to use and conceptually easier to understand' (MOORE and WILSON 1992).

Crop cover factor C

In USLE the *C* factor was taken from tables covering the most frequently encountered combinations of rotation, crop, and management practice, these being derived primarily from field plot experiments. In RUSLE the *C* values are calculated from four sub-factors which can be separately evaluated. These are:

- *PLU*, the prior land use factor, which accounts for the amounts of biomass in the soil using a residual decay model, and the effect of previous tillage practices on consolidation.
- *CC*, the canopy sub-factor, reflects the effectiveness of vegetative canopy reducing the energy of rainfall striking the soil surface. It is similar in effect to the relation between erosion and percentage cover used in the SLEMSA equation (section 7.2.4 and figure 7.2), and also shown in figure 11.1. It retains the *Handbook 537* allowance for the effect of the height of the canopy, since a high canopy allows throughfall rain to pick up more kinetic energy than a low canopy.
- *SC*, the surface cover factor, assesses the effect of mulch or other ground cover, and this interacts with
- *SR* the surface roughness factor.

The crop cover factor *C* is calculated over 15-day time periods, and combined with the 15-day values of *R* and *K*. One benefit of the new sub-factor method is that it can be applied to crops for which there are no experimental data.

Support practice factor P

This was the least satisfactory of the factors in USLE, in that it was very crude compared with the other factors, and much effort and research has gone into refining it for RUSLE. It still has the same basic function of allowing for the effects of practices which change the flow pattern or the rate or the quantity of surface runoff. For cultivated land these are mainly tillage practices on the contour, stripcropping, terracing, and sub-surface drainage. It does not include improved tillage practices such as no-till and other forms of conservation tillage, or crop residue management as these are considered in the *C* factor.

Because support practices are usually used in combinations, such as

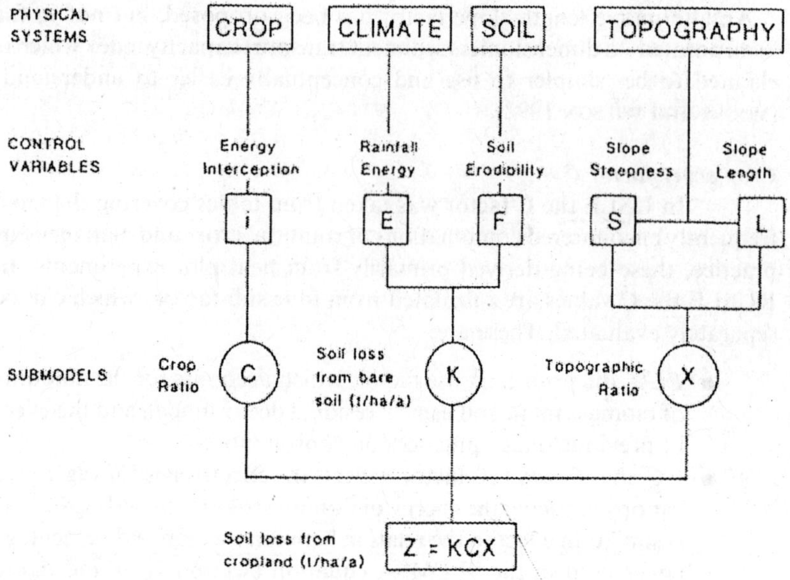

Figure 7.2 The structure of SLEMSA (ELWELL 1981)

contour tillage and stripcropping, an overall *P* factor is computed as a product of sub-factors. The main sub-factors are the effect of tillage on the true contour, or off-grade, the height of ridges and the risk of overtopping, the effect of grade of terraces on runoff and on soil deposition, and the reduction of soil movement and runoff by various forms of stripcropping.

Operation

RUSLE is still, like USLE, an empirical model based on experimental data. For USLE the analysis was done before computers, using machine sorting of punched cards at a single centre. For RUSLE the analysis of a much larger database used mainframe computers operated by a nationwide network of researchers, but a primary requirement was that the end product should be user-friendly, and suitable for operation by field staff and extension agents on personal computers.

7.2.5 *Other empirical models derived from USLE*

Some of the variations on USLE are to cater for local conditions which are too different from the USA situation, such as steepland, tropical crops, or soil properties outside the US range (such as organic matter). Other variations are in order to get local factors more simply than in USLE, or from a smaller databank.

No other country has a database from experimental plots comparable with that of the USA which has the results from more than 10 000 plot years. Australia and Zimbabwe each have a significant base from long-running national research programmes, for example 50 plots for 20 years. Many other countries have some runoff plots, and regional networks are stimulating many more. However, not all these experiments were designed to produce data for prediction equations, and the quality of the data often leaves much to be desired.

The Soil Loss Estimation Model for Southern Africa (SLEMSA) The question therefore is how to make use of limited data in a system that allows progressive improvement as more data are acquired? SLEMSA is a good example (ELWELL 1981, ELWELL and STOCKING 1982). The framework of the model in figure 7.2 shows that it closely follows that of the USLE.

The basic USLE equation in real units is $A = RK$, with modifying ratios C,L,S, and P. The basic SLEMSA equation in real units is $Z = K$ (a combination of rainfall E and erodibility F), with modifying ratios C derived from cover, and X from L and S.

Both models estimate long-term average annual soil loss, and in both the combination of cover and rainfall energy can be done for each of the smaller periods of time within the growing season.

The difference between the two models are that in SLEMSA

- **The P factor** of USLE is left out because it is felt that the effect of local conservation practices can be allowed for in factors L or S within the topography system, or within erodibility F in the soil system.
- The other factors are quantified by methods which are simpler to calculate or require less data.

R in USLE is replaced by E in SLEMSA, and is a measure of the total annual kinetic energy of the rainfall, which is easier to calculate from rainfall records than EI.

C in USLE is replaced by a different C in SLEMSA and is determined from 'i' the density of crop cover which is measured in the field at 10-day intervals over the 180-day growing season (STOCKING 1988). C is expressed as a ratio of the soil lost from a cropped plot to that lost from a bare fallow. SLEMSA can be used to estimate the soil loss from range land using a slightly different sub-model to relate *C* to *i*.

K in USLE is replaced by *F* in SLEMSA which has units of t/ha/a and is based on soil type.

LS in USLE is replaced by *X* in SLEMSA calculated in very similar manner but with slightly different equations.

Some other derivations from USLE

SOILOSS is a model developed in New South Wales, Australia which uses the same format and factors, as USLE, but uses local experience to vary the estimation of the factors, *R*, *K*, and *C*.

Erosivity factor EI Long records of rainfall including a six-year study of drop size distribution, energy, and intensity, made it possible to establish a correlation between EI and the two-year six-hour rainfall of the same form as has been established in the United States as discussed in section 4.2.1.

It is suggested that EI could profitably be replaced by a more process-based index QE_A, where Q is the runoff rate and E_A is the rate of expenditure of rainfall kinetic energy (KINNELL 1985). Alternatively I_X, the excess rainfall rate, can be used as a surrogate for Q. This change may be incorporated in later versions of SOILOSS.

The erodibility factor K The USLE nomograph for predicting erodibility (figure 5.1) uses particle size distributions derived from laboratory analysis after shaking a soil sample with a chemical dispersant. Several workers have felt that this is more aggressive disaggregation than takes place in the field under raindrop impact and runoff flow. LOCH and ROSEWELL (1992) show that feeding into the nomograph the particle size distribution after shaking with water alone improved the correlation with field measurements of *K* available for six soils in eastern Australia.

Crop management factor C One of the new introductions in RUSLE (section 7.2.3) is the introduction of sub-factors for prior land use, crop canopy cover, crop residue cover, and surface roughness. These have been determined for a range of agricultural crops grown in New South Wales using simple plant growth, tillage, and residue decay models which have been calibrated with local data (EDWARDS and ROSEWELL 1990).

New South Wales has a database of 4 500 plot years over periods of up to 40 years which has been used to validate the model and to make adjustments, for example making an allowance for summer weed growth during fallow periods in a winter wheat system. Like RUSLE, a computer programme and instruction manual are commercially available and in regular use by field staff and in higher education (ROSEWELL and EDWARDS 1988). The soil loss package was primarily developed for use on the crop lands of New South Wales and a two-year national project will expand its use to urban land, pasture and rangeland, and for croplands in other States, and will continue to incorporate appropriate new algorithms used in RUSLE (ROSEWELL 1992).

Applying USLE to forest land

The usefulness of USLE in its original purpose of estimating erosion from cropland has led to its being extended to other situations, sometimes successfully, sometimes less so. The technique is to define how another situation differs from the crop situation and to quantify the differences using the sub-factor approach. For example, applying USLE to forest land, an obvious difference is that the canopy is firstly not regularly distributed as it is in row crops, and secondly the height of tree canopy is beyond the range of crop heights considered in USLE. There are other differences: as a result of the decay of leaf litter on the forest floor, the organic matter content may be above the 4% limit of the erodibility nomogram: a dense fine root mat is usually present in the top 25 mm of most forest soils, reducing the erodibility. Sub-factor relationships for these and other differences have been developed using experimental data, experience, and field observations. The relationships have been validated and where necessary modified by comparing estimated soil loss values with measured values. Multiplying the values of all the sub-factors gives a combined value for *C*, the crop management factor in USLE (DISSMEYER and FOSTER 1980, 1985). The sub-factor approach is used in RUSLE to assist the extension to rangeland by considering the effect of increased infiltration and surface roughness, interrelated with slope steepness.

SCUAF – Soil Changes Under Agroforestry

The USLE may also be used as a component in models where erosion is just one of several factors which affect a land use system, and SCUAF is an example of this. The objectives are:

- To predict the effects on soils of specified agroforestry systems in given environments.
- To show what data are needed to make such predictions.
- To use predictions in the design of systems for agroforestry research.
- To indicate what advances in knowledge are needed in order to improve the accuracy of the predictions (YOUNG 1989).

The model operates on a year-by-year time base, and has two main compartments, a plant compartment and a soil compartment. The soil compartment includes a carbon cycle, a nitrogen cycle, and an estimate of erosion from USLE, using the *C* factor as modified by Dissmeyer and Foster. A phosphorus cycle could be added later.

For the Mark 1 model there are four outputs:

- Changes in soil humus carbon.
- Changes in soil erosion.
- Changes in plant biomass production as affected by soil.
- Changes in harvest.

In the Mark 2 model there are additional outputs of changes in organic and mineral nitrogen (YOUNG and MURAYA 1988).

INDEROSI – an erosion estimation model from Indonesia

In the planning and evaluation of erosion control systems it may be useful to be able to measure relative changes in erosion, without trying to estimate absolute values of soil loss. A model is being developed for this purpose in Indonesia, called INDEROSI (GNAGEY 1991). The model assumes that the real change in erosion over time on any particular site is proportional to change in the crop management factor *C* and the support practice *P*. The estimation of *C* and *P* has been simplified so that data can be quickly collected from a large number of small plots which comprise a site of perhaps 10 ha. A computer programme calculates the annual cover factor and treatment factor for each plot, and then combines these, weighted according to the area of each plot, into an erosion index ($C \times P$) for the site, which is fed into the USLE. This exercise is repeated annually to give a qualitative indication of the change in erosion year on year, or before and after project.

7.3 Process-based or physically-based models

The creation of complicated models based on physical processes appears to have a fascination for academics and their graduate students. There are many models – some at early stages of development, some being validated, a few being used, and a very few being used by other people in addition to their originators. We will not attempt to review them all, only a selection chosen either because they seem likely to be adopted such as WEPP, or because they introduce important new concepts. Other references are included in *Further reading*.

The starting point for physically-based models was MEYER and WISCHMEIER (1969) who defined four basic erosion processes: detachment by raindrop splash, transportation by raindrop splash, detachment by surface runoff, and transportation by surface runoff. This was developed by FOSTER and MEYER (1975) who added the concept of separating rill from inter-rill erosion, and led to a basic equation by FOSTER *et al* (1977). Legislation in the United States controlling water quality (particularly section 208 of Public Law 92–500) has resulted in a need to combine soil loss into the much wider picture of surface runoff and pollution from all sources including sediment. A working model of the whole of surface hydrology CREAMS includes a sub-model for erosion (FOSTER *et al* 1981b). A review of the development of modelling up to the present time is given by LANE *et al* (1992).

7.3.1 EUROSEM (European Soil Erosion Model)

This is a cooperative model-building project of the European Community with collaborating scientists in seven countries and purpose-built runoff plots for validating the model in England and in Spain (MORGAN *et al* 1992, QUINTON and MORGAN 1993).

Objectives

The design requirements include:

- Assessment of erosion on field and catchment scale.
- Assessment of pollution by sediment or solutes to water bodies.
- To operate on an event basis.
- To be useful as a design tool for selecting soil protection measures.

Elements of the model

Rainfall is assessed by the amount of interception by vegetative cover and the proportions of intercepted rainfall reaching the ground as stemflow or as leaf drip.

Soil surface condition is based on quantifying roughness to assess surface storage.

Runoff generation is separated into two components: surface depression and surface flow. The hydrological model simulates runoff as both Hortonian and saturation flow expressed as a depth.

Soil detachment by raindrop impact is assessed from the kinetic energy of the throughfall and separately the energy of leaf drip from the canopy.

Soil detachment by runoff is assessed from a comparison of the estimated velocity of runoff compared with the critical velocity required to detach soil particles, modelled as a function of grain shear velocity.

Transport capacity is modelled as a function of stream power defined as the product of flow velocity and slope.

Finally a comparison is made of the availability of detached soil for transport, and the transport capacity of flow, done separately for flow in depressions and for surface flow. In each case there can be net erosion or deposition.

7.3.2 CREAMS (Chemicals, Runoff, and Erosion from Agricultural Management Systems)

Objectives

CREAMS is not a predictive model in absolute terms, but allows the comparison of the effect of different practices on an event basis. It is probably the most widely tested of process-based models, and has been used mainly to compute the loss of pollutants from field-scale agricultural areas, but the erosion component of the model can be modelled separately. Originally it was intended to operate on areas up to 40 ha, but it can be stretched to 400 ha (LANE *et al* 1992). When used for a comparison of alternative practices it does not need to be validated or calibrated. It may be applied also to range land. The definitive explanation of the model is a huge report edited by KNISEL (USDA 1980) with a more manageable summary by FOSTER *et al* (1981b). A revised version CREAMS 2 is also referred to as OPUS (SMITH and KNISEL 1985).

Features

There are three components of the model: hydrological, erosion, and chemistry.

The *hydrological component* estimates runoff volume and peak rate, infiltration, evapotransportation and percolation on a daily basis.

The *erosion component* is partly physically based and partly empirical. Erosion and sediment yield are estimated in three phases.

- Overland flow, where detachment by rill and inter-rill flow is compared with the transport capacity to estimate rates of erosion and deposition. This phase used parameters of erosivity and erodibility from the USLE.
- Channel flow, where scour erosion is estimated in larger channels before leaving the field (called *ephemeral gully erosion*).
- Deposition, where flow velocity is decreased by change of slope or ponding caused by a physical structure such as a channel terrace.

The *chemistry element* considers plant nutrients and pesticides and gives estimates of storm loads which are computed from the average concentrations of absorbed and dissolved chemicals in the runoff and sediment.

7.3.3 ANSWERS (Areal Non-point Source Watershed Environment Response Simulation)

Objective

The purpose of this model is to assess sediment yield from the whole watershed (a non-point source) as opposed to that from field-size areas (a point source), and to assess the cost effectiveness of possible land use treatments within the watershed. The model can only be operated on a mainframe computer because it makes use of a huge database compiled from sources such as topographic maps, soil surveys, crop and management surveys. When the model was created in 1980 this required a major effort which would be much easier nowadays using GIS techniques to digitize this kind of data.

Components

The watershed is divided into elements, usually of approximately 4 ha in which the input parameters can be assumed to be uniform. Each element is analysed separately using three sub-models: hydrologic, sediment detachment/transport, and a number of routing components which describe the movement of water in overland, sub-surface, and channel flow phases (BEASLEY *et al* 1980). The erosion sub-model is basically the usual comparison of detachment capacity with transport capacity, and is partly empirical, using some of the factor relationships from USLE. A later version (PARK *et al* 1982) replaces this with a process-based sub-model for erosion and sediment transport. When an element has a defined channel, the output (sediment yield and runoff) is passed as an input to the next element downstream in the channel. In elements where there is no defined channel the output is passed to adjoining elements as required by the steepest slope, and if an element slopes towards more than one adjoining element the output can be divided and apportioned appropriately.

7.3.4 WEPP (Water Erosion Prediction Project)

Following a workshop in 1985 the USDA initiated a ten-year research project in cooperation with other federal agencies including the Agricultural Research Service (ARS), Soil Conservation Service (SCS) the Forest Service, and the Bureau of Land Management (BLM). Thre phases were contemplated, phase one from 1985 to 1989 for developing the technology to arrive at a prototype for testing, phase two from 1989 to 1992 for testing and refinement of the model, and phase three, 1992 to 1995, for implementation and training in the use of the model (LAFLEN *et al* 1991).

Purpose

The objective is to develop a new generation of erosion prediction technology, based on current understanding of erosion processes, and that the technology should be applied at the same level and for the same uses as current USLE technology, and in addition be applicable to a wider range of scales and land use possibilities.

Features

WEPP operates on a daily timestep and a basic requirement was that it should be suitable for a wide range of personal computers. There are three versions, which are in effect a question of scale.

The *profile version* replaces USLE as an estimator of erosion on a uniform hillslope, with the addition of an element for deposition.

The *watershed version* extends to a field size watershed such as a terraced field. It includes the profile version which may be applied to more than one profile area to estimate the delivery of sediment to channels, and adds the effects in small channels (ephemeral gully erosion) and possible deposition by ponding, to arrive at delivery of sediment at the outlet from the watershed.

The *grid version* is for areas whose boundaries do not coincide with watershed boundaries, and which can be broken into separate elements, to each of which the profile version can be applied. It then handles movement from element to element (as in ANSWERS), and delivery from more than one discharge point.

Processes

The erosion processes considered are detachment, transport, and deposition, considered separately in rill and inter-rill erosion. *Detachment* in rill erosion is estimated from linear functions of excess hydraulic shear. The *Yalin equation* is used for sediment *transport* in channels and rills. *Deposition* in ponding areas is treated as in the CREAMS model.

Hydraulic processes include climate, infiltration, and the kinematic routing of runoff.

Plant growth and residue processes are concerned with growth and decay both above and below ground.

Water-use processes use information from components for climate, plant growth, and infiltration to estimate daily potential and actual evapotransportation.

Hydraulic processes use data on surface runoff and hydraulic roughness to compute hydraulic shear forces in rills.

The *soil component* considers the effects of tillage, weathering, consolidation, and rainfall on soil variables. The research plan includes a major effort to develop the prediction of erodibility from soil properties beyond the estimation of the erodibility factor *K* in USLE.

The development of WEPP is running approximately to plan and it is expected that it should be fully operational during 1995.

7.4 Productivity models

Models to estimate the loss of productivity are relatively new as compared to models for estimating soil loss, but are becoming increasingly needed as the loss of productivity is seen to be more significant than the volume or weight of eroded soil as discussed in section 2.2.3. Most loss of productivity models start by estimating the amount of soil loss, and then as an additional stage assess the effect of this loss on productivity. This is not entirely satisfactory as it does not account for loss of nutrients in runoff, nor for loss of productivity through degradation of the physical structure such as compaction or reduced permeability.

7.4.1 EPIC (Erosion Productivity Impact Calculator)

The most comprehensive productivity model is EPIC which is a combination of empirical and physically based components. A huge input of data is required, but all from sources readily available in the USA. A mainframe computer is used with a daily time-step and 53 sub routines. It is applied to fairly small areas, about 1 ha, which are assumed to be spatially homogeneous, but the soil profile is divided into up to 10 layers whose depth and soil properties can vary. Surface runoff is predicted in a procedure similar to that of CREAMS, and other components of the hydrologic phase are weather, percolation, drainage, and evapotransportation. Erosion estimates are based on a variation of USLE, and the plant nutrients considered are nitrogen and phosphorus. There is a single growth model applicable to all major field crops grown in North America, and a tillage component which mixes nutrient and crop residue within the plough depth. The model can estimate changes in production over periods of up to 50 years, and is expected to become an important tool for policy studies, and planning at national and programme level (WILLIAMS 1985, WILLIAMS *et al* 1983 and 1990).

7.4.2 *Productivity Index model*

Another approach is the Productivity Index model (PIERCE *et al* 1983), designed to estimate long-term effects of erosion. Like EPIC a large input of data is taken from existing databases, in this case *SOILS-5* which contains physical and chemical properties, crop yields, and land capability and limitations for each soil series in the USA, and the NRI database (Natural Resources Inventory compiled in 1977) which contains information on size, land use, ownership, erosion and other relevant factors for all soil mapping units in the USA. The Productivity Index considers the sufficiency or adequacy of the attributes of soil productivity in separate layers, weighted for proportion of crop roots in each layer. The attributes are available water storage, aeration, bulk density, pH, and electrical conductivity. Nutrients are not included, on the assumption that deficiencies can be corrected by the application of fertilizers.

The effect on erosion rates, as reported in the NRI survey, for 20, 50, and 100 years, is calculated as the soil surface is progressively eroded. As would be expected, the effect varies greatly on different soil types. Some are intolerant of erosion and quickly lose productivity, others such as deep loess soils are less affected. The purpose of the model is to identify the risk of long-term damage, so that land use policies and planning can be more effectively targeted.

Models which, like EPIC and PI, require a huge database are not applicable in developing countries, so a number of attempts have been made to build simpler productivity models. The concept of sufficiency of essential productivity attributes, as used in PI, was used by ELWELL and STOCKING (1984) to develop the concept of *soil life* or *residual suitability*, i.e. for how long can the soil produce at an acceptable level of yield as it degrades through erosion. What constitutes an acceptable yield depends on socio-economic criteria, and a simple model, based on the observed relationship between soil loss and yield, was developed and tested for range land in Botswana and for cereal cropping in Sierra Leone (BIOT *et al* 1989). In a further development, THEPROM (Theoretical Erosion Productivity Model, BIOT 1990) assumes that the critical factor determining productivity is Available Water Storage Capacity (AWSC), which is the product of the effective rooting depth of the soil, and its available water storage.

7.5 Watershed models

When the objective is to understand the off-site effects of erosion, such as the chemical or physical pollution of streams and rivers, then assessment of in-field movement by models such as USLE do not help because they do not attempt to estimate how much soil or sediment is actually leaving the field. Some models add on a delivery ratio to the end of USLE (*eg* MUSLE), others build an estimate of sediment deposition into the erosion component (*eg* CREAMS). But this still leaves the question of off-farm sediment production or deposition, such as gullying and streambank erosion, or erosion which is associated with non-agricultural features such as roads and urban areas. Some watershed models approach this problem by dividing the watershed into a large number of small elements each of which is analysed separately and can include off-farm factors. Other watershed models are hydrologic models with a component estimating sediment yield, and work on this topic by USDA-ARS is reviewed by RENARD (1993). These are often described as 'non-point source' models and may have a structure quite different from that of USLE, for example the rainfall effect may be represented by a factor calculated from the volume of runoff and peak rate instead of the R factor based on rainfall energy (WILLIAMS 1975).

References

BARNETT, A. P. 1958 How Intense Rainfall Affects Runoff and Soil Erosion, *Agricultural Engineering*, 39, 11, 703–707, 711

BEASLEY, D. B., L. F. HUGGINS and E. J. MONKE 1990 ANSWERS: A model for watershed planning, *Transactions of the Association of American Engineers*, 23, 4, 938–944

BIOT, Y. 1990 THEPROM: an erosion-productivity model, in *Soil Erosion on Agricultural Land*, 465–479, edited by J. Boardman, I. Foster and J. Dearing, Wiley, Chichester, Sussex

BIOT, Y., M. SESSAY and M. STOCKING 1989 Assessing the sustainability of agricultural land in Botswana and Sierra Leone, *Land Degradation and Rehabilitation*, 1, 263–278

BROWNING, G. M., C. L. PARISH and J. A. GLASS 1947 A method for determining the use and limitation of rotation and conservation practices in control of soil erosion in Iowa, *Journal of the American Society of Agronomy*, 39, 65–73

DISSMEYER, G. E. and G. R. FOSTER 1980 *A guide for predicting sheet and rill erosion on forest land*, Technical Publication SA-TP11, USDA Forest Service, Atlanta, Georgia

DISSMEYER, G. E. and G. R. FOSTER 1985 Modifying the Universal Soil Loss Equation for forest land, chapter 45, in *Soil Erosion and Conservation*, edited by S. A. El-Swaify, W. C. Moldenhauer, and A. Lo, Soil Conservation Society of America, Ankeny, Iowa

EDWARDS, K. and C. J. ROSEWELL 1990 Evaluation of alternative land management and cropping practices for soil conservation, *Soil Use and Management*, 6, 120–124

ELWELL, H. A. 1981 A soil loss estimation technique for southern Africa, in *Soil conservation: Problems and Prospects*, 281–292, edited by R. P. C. Morgan, Wiley, Chichester, Sussex

ELWELL, H. A. and M. STOCKING 1982 Developing a simple yet practical method of soil-loss estimation, *Tropical Agriculture* (Trinidad), 59, 1, 43–47

ELWELL, H. A. and M. STOCKING 1984 Estimating soil life-span for conservation planning, *Tropical Agriculture*, 62, 2, 148–150

FOSTER, G. R. 1990 Process-based modelling of soil erosion by water on agricultural land, chapter 28, in *Soil Erosion on Agricultural Land*, edited by J. Boardman, D. L. Foster and J. A. Dearing, Wiley, Chichester, Sussex

FOSTER, G. R. and L. D. MEYER 1975 Mathematical Simulation of Upland Erosion by Fundamental Mechanics, in *Present and Prospective Technology for predicting Sediment Yields and Sources*, United States Department of Agriculture, ARS-S-40, 190–207

FOSTER, G. R., L. D. MEYER and C. A. ONSTAD 1977 An erosion equation derived from basic erosion principles, *Transactions of American Society of Agricultural Engineers*, 20, 678–682

FOSTER, G. R., D. K. MCCOOL, K. D. RENARD and W. C. MOLDENHAUER 1981a Conversion of USLE to SI Units, *Journal of Soil and Water Conservation*, 36, 6, 355–359

FOSTER, G. R., L. J. LANE, J. D. NOWLIN, J. M. LAFLEN and R. A. YOUNG 1981b Estimating Erosion and Sediment Yield on Field-Sized Areas, *Transactions of American Society of Agricultural Engineers*, 24, 1253–1262

FOSTER, G. R., W. C. MOLDENHAUER and W. H. WISCHMEIER 1982 Transferability of US technology for prediction and control of erosion in the tropics, in *Soil Erosion and Conservation in the Tropics*, Special Publication 43, American Society of Agronomy

GNAGEY, R. W. 1991 *Calculating erosion reduction with INDEROSI*, Paper presented at International Workshop for Sustainable Hillslope Farming, Solo, Indonesia, March 1991

GREER, J. D. 1971 Effects of Excessive-rate Rainstorms on Erosion, *Journal of Soil and Water Conservation*, 26, 196–197

HUDSON, N. W. 1957 Erosion Control Research, Progress Report on Experiments at Henderson Research Station 1953–1956, *Rhodesian Agriculture Journal*, 54, 4, 297–323

KINELL, P. I. A. 1985 Runoff effects on the efficiency of raindrop kinetic energy in sheet erosion, chapter 37, in *Soil Erosion and Conservation*, edited by S. A. El-Swaify, W. C. Moldenhauer and A. Lo, Soil Conservation Society of America, Ankeny, Iowa

LAFLEN, J. M., L. J. LANE and G. R. FOSTER 1991 WEPP – A new generation of erosion prediction technology, *Journal of Soil and Water Conservation*, 46, 1, 34–38

LANE, L. J., K. G. RENARD, G. R. FOSTER and J. M. LAFLEN 1992 Development and Application of Modern Soil Loss Prediction Technology – The USDA Experience, *Australian Journal of Soil Resources*, 30, 893–912

LITTLE, J. M. 1940 *Erosional Topography and Erosion*, Carlisle, San Francisco

LOCH, R. J. and C. J. ROSEWELL 1992 Laboratory Methods for measurement of soil erodibilities (K factors) for the Universal Soil Loss Equation, *Australian Journal of Soil Research*, 30, 2, 233–248

MEYER, L. D. and W. H. WISCHMEIER 1969 Mathematical simulation of the process of soil erosion by water, *Transactions of American Society of Agricultural Engineers*, 12, 6, 754–758, 762

MEYER, L. D. and C. HARMON 1985 Sediment losses from cropland furrows of different gradients, *Transactions of American Society of Agricultural Engineers*, 28, 2, 448–453, 461

MOORE, I. D. and J. P. WILSON 1992 Length-slope factors for the Revised Universal Soil Loss Equation: simplified method of estimation, *Journal of Soil and Water Conservation*, 47, 5, 423–428

MORGAN, R. P. C., J. N. QUINTON and R. J. RICKSON 1992 A soil erosion prediction model for the European Community, chapter 12, in *Soil Conservation for Survival*, edited by H. Hurni and K. Tato, Soil and Water Conservation Society, Ankeny, Iowa

MUSGRAVE, G. W. 1947 Quantitative Evaluation of Factors in Water Erosion – a First Approximation, *Journal of Soil and Water Conservation*, 2, 133–138

PARK, S. W., J. K. MITCHELL and J. N. SCARBOROUGH 1982 Soil erosion simulation on small watersheds: A Modified ANSWERS model, *Transactions of American Society of Agricultural Engineers*, 25, 6, 1581–1588

PIERCE, F. J., W. E. LARSON, R. H. DOWDY and W. A. P. GRAHAM 1983 Productivity of Soils: Assessing long-term changes due to erosion, *Journal of Soil and Water Conservation*, 38, 1, 39–51

QUINTON, J. N. and R. P. C. MORGAN 1993 *Description of the European Soil Erosion Model and an example of its validation*, Paper presented at International Workshop on soil erosion processes on steep lands;

evaluation and modelling, Merida, Venezuela, May 1993

RENARD, K. G. 1977 Erosion research and mathematical modelling, in *Erosion research techniques, erodibility and sediment delivery*, edited by T. J. Toy, Geo Abstracts, Norwich

RENARD, K. G. 1993 Past, Present, and Future Hydrologic Modeling in ARS, in *Proceedings of Federal Interagency Workshop on Hydrologic Modeling Demands for the 90s*, edited by J. S. Burton, US Geological Survey Water Resources Investigation Report 93–4018

RENARD, K. G., G. R. FOSTER, G. A. WEESIES and J. P. PORTER 1991 RUSLE – Revised Universal Soil Loss Equation, *Journal of Soil and Water Conservation*, 46, 1, 30–33

ROSEWELL, C. J. 1992 The Development of Land Protection Technology in Australia, in *People Protecting their Land*, edited by P. G. Haskins and B. M. Murphy, Department of Conservation and Land Management, Sydney, New South Wales

ROSEWELL, C. J. and K. EDWARDS 1988 SOILOSS: A program to assist in the selection of Management Practices to Reduce Erosion, *Technical Manual 11*, Soil Conservation Service of New South Wales

SMITH, D. D. 1941 Interpretation of Soil Conservation Data for Field Use, *Agricultural Engineering*, 22, 173–175

SMITH, R. E. and W. G. KNISEL 1985 Summary of methodology in the CREAMS2 Model, in *Proceedings of Natural Resources Modeling Symposium, Pingree Park, Colorado*, edited by D. G. Decoursey, ARS 30 33–66, USDA, Washington DC

STOCKING, M. A. 1988 Assessing vegetative cover and management effects, in *Soil Erosion Research Methods*, 7, 141, edited by R. Lal, Soil and Water Conservation Society, Ankeny, Iowa

USDA-ARS 1994 Predicting Soil Erosion by Water: A guide to conservation planning with the Revised Universal Soil Loss Equation, *Agricultural Handbook 703*, USDA ARS, Washington DC

USDA-SEA 1980 A Field Scale Model for Chemicals, Runoff, and Erosion from Agricultural Management Systems, *Conservation Research Report 26*, edited by W. G. Knisel in 3 volumes, 643

VANESLANDE, A., R. ROUSSEAU, R. LAL, D. GABRIELS and B. S. GHUMAN 1984 Testing the applicability of a soil erodibility nomogram for some tropical soils, in *Challenges in African Hydrology and Water Resources*, edited by D. E. Walling, S. S. D. Foster and P. Wurzel, IAHS publication 144, 463–473

WILLIAMS, J. R. 1975 Sediment-yield prediction with Universal Equation using runoff energy factor, in *Present and Prospective*

Technology for Predicting Sediment Yields and Sources, ARS-S-40, USDA, Washington DC, 244–252

WILLIAMS, J. R. 1985 The physical components of the EPIC model, in *Soil Erosion and Conservation*, edited by S. A. El-Swaify, W. C. Moldenhauer and A. Lo, Soil Conservation Society of America, Ankeny, Iowa

WILLIAMS, J. R., K. G. RENARD and P. T. DYKE 1983 A new method for assessing the effect of erosion on predictability – the EPIC model, *Journal of Soil and Water Conservation*, 38, 381–3

WILLIAMS, J. R., A. N. SHARPLEY and D. TAYLOR 1990 Assessing the Impact of Erosion on Soil Productivity using the EPIC model, in *Soil Erosion on Agricultural Land*, edited by Boardman, Foster and Dearing, Wiley, Chichester, Sussex

WISCHMEIER, W. H. 1962 Storms and Soil Conservation, *Journal of Soil and Water Conservation*, 17, 2, 55–59

WISCHMEIER, W. H. 1976 Use and Misuse of the Universal Soil Loss Equation, *Journal of Soil and Water Conservation*, 31, 1, 5–9

WISCHMEIER, W. H., D. D. SMITH and R. E. UHLAND 1958 Evaluation of Factors in the Soil Loss Equation, *Agricultural Engineering*, 39, 8, 458–462, 474

WISCHMEIER, W. H. and D. D. SMITH 1978 Predicting rainfall erosion losses, *United States Department of Agriculture, Agricultural Handbook 537*, and supplement of 1981, Washington DC

YOUNG, A. 1989 Modelling soil changes under Agroforestry, chapter 15, in *Agroforestry for Soil Conservation*, C. A. B. International, Wallingford, Oxford

YOUNG, A. and P. MURAYA 1988 Soil changes under agroforestry (SCUAF), in *Land Conservation for Future Generations*, edited by Sanarn Rimwanich, Vol 1, 655–667, Department of Land Development, Bangkok, Thailand

ZINGG, A. W. 1940 Degree and Length of Land Slope as it affects Soil Loss in Runoff, *Agricultural Engineering*, 21, 59

8

Erosion research methods

8.1 The purpose of erosion research

8.1.1 The need for research

A soil conservation programme needs to be based on factual information about rates and quantities of erosion. For example some land obviously needs to be protected by mechanical works such as channel terraces, some land equally obviously does not need any protection, and a great deal of land lies somewhere between the two extremes. At some point the terraces become unnecessary or uneconomic, but a logical decision on where to draw the line can only be made if data are available on how much soil erosion is occurring and how much it would be reduced by terraces. Similarly, the effect of channel terraces should be compared with the effect of alternative measures, and the effect should also be measured on the different soils and crops which are likely to be encountered.

Nowadays there are excellent opportunities for research workers to exchange the results of their work. The national and international journals are packed with accounts and results of research reports in their proceedings (as listed in *Further reading*). Some of the reported research results can be usefully applied in countries other than where the research was done, but care is required in any transfer of technology. If the local conditions of soil, climate, or land use, are different, the results may not apply. However, research in other situations can often serve as a pilot trial to show which factors should be studied locally. Even when local conditions are so different that the results do not apply at all, the techniques and apparatus developed by other workers can still be helpful.

8.1.2 Defining the objectives

Before deciding on a research programme, it is important to define the objectives very clearly. If the purpose is to obtain a practical answer to a practical problem, such as which of two popular rotations is more subject to erosion, then the experimental procedures will be very different from those required to establish long-term fundamentals such as the erosive power of the local rainfall. The scale of the experiments also needs to be considered, for analytical studies of detailed points might be done on small plots or in the laboratory, but practical farming operations can only be tested on plots large enough for farm-scale practices. The required accuracy of the results must also be defined, for a precise and accurate technique will be wasted if the question is only 'which of two alternatives is better?'

8.1.3 Changes in direction

In recent years there have been a number of changes in the emphasis on different aspects of the problems to be tackled by research.

The first is the change of scale. Historically, erosion research started in the United States where the farming pattern is large mechanized arable farms, and the same style of research was continued in those countries with a similar farming pattern, such as some former Colonial territories in Africa, and in Australia, and in parts of the Soviet Union. Today it is the research needs of the small-scale subsistence farmer which deserve attention and are increasingly being addressed. This change is sometimes associated with more use of on-farm trials and adaptive research by farmers. The problem of on-farm research is that it nearly always suffers from the problem that the variables cannot be separated and controlled. An example is the studies made in Kenya which were set up to check whether a conservation package programme increased the yield of maize on fields where the programme had been applied. Data collected from farmers' fields did suggest that there was an increase, but the package consisted of building earth terraces, using improved seed, using more fertilizer, and generally improved cultivation practices, and the effect of each of these factors could not be separately evaluated. It could be that the improved seed would have given the increase without the cost of additional fertilizer, or that the increased use of fertilizer could have given the improved yield if applied to the traditional maize varieties. If the effect of the terracing was merely to improve the soil moisture availability, that might have been achieved by simpler methods. Controlled experiments are needed to separate the variables, and on-station is the proper place. Focusing attention on the needs of the small-scale farmer has led to research into traditional farming and soil conservation practices which are discussed in section 9.3.

The second major change in direction is the growing interest in soil erosion as the cause of loss of productivity as opposed to loss of soil, so the focus is moving towards measuring the *effect* of erosion, rather than the *amount*. Unfortunately experiments set up to measure quantities and rates of soil loss seldom produced the additional information necessary to assess the effect on productivity, and so a whole new series of research projects has been established to provide information.

The third change in research direction is towards studies of the degradation of marginal lands which are being increasingly used as a result of population pressures. The starting point for most major research programmes has been how cultivated productive land could be protected, but now we need to know more about land which was previously considered unusable because it was too steep, or semi-arid, or had problems of salinity or waterlogging.

8.1.4 Using research results

There is a place for academic research in laboratories and universities on subjects such as the mechanics of erosion processes and the hydraulics of sediment movement because they are the basic building blocks for our understanding of erosion, but we are here more concerned with research which will produce information which can be directly applied to the control of erosion. A recent FAO publication on the design and operation of field experiments (HUDSON 1993) lists some do's and don'ts which can be summarized as:

- A small amount of reliable information is more useful than a large amount of information which cannot be used because it is unreliable. There is a temptation to try to answer too many questions by including too many variables or too many levels of each variable.

- There is no justification for the idea that any information is better than none. This is sometimes used to justify quick look-see trials, but using inaccurate or unreliable data is much more likely to cause problems than to improve matters.

- The research objective must match the time available. Long-term trends which require the collection of data over many years cannot be fitted into the timespan of a student research grant.

- It may sometimes be possible to extrapolate short-term results if longer records are available, for example rainfall or streamflow records. From long records it may be possible to assess how typical or untypical the short-term records were, but on the whole a sound principle in all science is to avoid extrapolating beyond the range of measured results.

• 'Before and after' experiments should be avoided because if there is a measurable difference between before the treatment and after, there is no way of telling whether the difference is directly caused by the treatment or arises from some other cause of change between the two sets of measurements, such as a different rainfall.

• Expressing differences as percentages should be avoided. For example, the summary of a research report says that the annual soil loss from the plot with treatment *A* was 80% less than from the plot with treatment *B*. This sounds impressive until one reads that the figures were 100 kg/ha and 20 kg/ha, so the real conclusion should have been that soil loss was neglible in both cases. Avoid also announcing a winner without giving the scores. A comparison was made between five different parameters for estimating the erosivity of rainfall, and the abstract says that number 4 was 'the best estimator'. The full report discloses that the correlation coefficient was above 0.9 for all five, with the 'winner' at 0.955. A more sensible conclusion would have been that all five were extremely effective, and the one to recommend should be whichever is simplest and easiest to use.

8.2 Reconnaissance studies of erosion

8.2.1 The accuracy of measurements

Quantitative measurements of erosion are most frequently and most accurately obtained from permanent plots, but this method is not always suitable. For example, exploratory trials to establish the order of magnitude of the loss do not warrant fixed plots. Another case is measuring localized erosion which cannot be fitted onto plots, such as erosion in gullies, or on roads, or that caused by local overgrazing. In such cases the order of accuracy required may be very different from that which is suitable for measuring the soil loss from different crop rotations.

There are two approaches to estimating soil movement. One can estimate how much has been lost from a site, or estimate how much has accumulated somewhere else. The accuracy of these two approaches is usually very different. If the soil eroded from a field plot is collected in tanks and weighed, then even a coarse measurement of the weight gives a precise estimate of the soil loss. For example, if the soil loss from a plot of 100 m^2 is measured to the nearest 0.1 kg, this corresponds to an estimate of soil loss to the nearest 10 kg/ha. In comparison, a direct measurement of the level of the soil surface is a very crude estimate. If the lowering of the soil surface is measured to the nearest millimetre, this corresponds with an estimate of the soil loss to the nearest 15 000 kg/ha, i.e. the precision is about 1 500 times less.

The main advantage of reconnaissance methods is that, because they are cheap and simple, many measurements can be made and so the results can be reliable and representative – which means they are believable and usable – more so than a single precise measurement which may not be representative. The direct measurement of changes in soil level is appropriate in the case of localized erosion where rates are high and the position of the erosion can be predicted, such as steep land which has been deforested, or cattle tracks on rangeland. It is usually not suitable for losses from arable land because the surface level is affected by cultivation and settlement, although short-term changes have been studied in potato furrows in Australia (MCFARLANE *et al* 1991). Changes can be measured in one dimension for surface level at a point, or in two dimensions to give a profile or cross-section, or in three dimensions for volumetric measurements of rills or gullies.

8.2.2 *Point measurements*

Individual measurements of change in level at a single point will vary widely, but if it is an inexpensive and simple method, and a large number of points can be sampled, then a usable estimate can result.

Erosion pins

This widely-used method consists of driving a pin into the soil so that the top of the pin gives a datum from which changes in the soil surface level can be measured. Alternatively called *pegs*, *spikes*, *stakes*, or *rods*, the pins can be of wood, iron, or any material which will not rot or decay and is readily and cheaply available. Off-cut lengths of round iron bars for reinforced concrete can usually be picked up for little cost from construction sites. In some developing countries iron or steel pins or nails might be stolen, in which case bamboo or reed canes cut locally might be more suitable (figure 8.1).

Some researchers slip a metal washer over the pin to give a better base from which to measure to the top of the pin. If there are likely to be cycles of erosion and deposition such as in a gully floor, the washer method may give useful additional information by falling to the lowest erosion level and being covered by any later deposition which can also be measured. On the other hand the presence of the washer may cause turbulence and scour, or it could reduce splash erosion and leave the washer sitting on a pedestal of soil. All these variations and possible causes of false readings have been

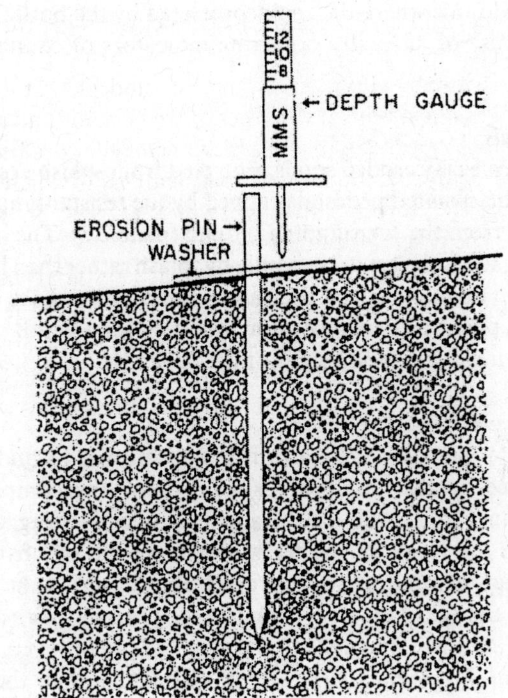

Figure 8.1 Erosion pins to measure change in surface level (FAO)

reported in the literature on the use of the pin method which is reviewed by HAIGH (1977).

Paint collars

An indication of large changes in level, for example in a stream bed or gully floor, can be obtained by painting a collar just above soil level round rocks, boulders, tree roots, fence posts, or anything firm and stable. Erosion reveals an unpainted band below the paint line, indicating the depth of soil removed. When painting the collar it is advisable to mask the soil with old newspaper as paint accidently sprayed or brushed onto the soil might make it less erodible.

Bottle tops

Another simple way to record the original level is to press bottle tops into the soil surface. The depth of subsequent erosion is shown by the height of the pedestals where the soil is protected by the bottle top. This leads us to the use of naturally occurring indicators of changes in soil surface level.

Pedestals

When an easily eroded soil is protected from splash erosion by a stone or tree root, isolated pedestals capped by the resistant material are left standing up from the surrounding ground (plate 2.3). The erosion of the surrounding soil is shown to be mainly by splash rather than by surface flow if there is no undercutting at the base of the pedestal. Like the bottle top method it is possible to deduce approximately what depth of soil has been eroded by measuring the height of the pedestals.

Tree mounds and tree roots

In arid or semi-arid climates it is not unusual to find that the surface under trees is raised in a gently sloping dome. In a comprehensive project in Tanzania from 1968–1972 Rapp and colleagues suggested that the mounds are the result of the trees protecting the soil from splash erosion while the surrounding soil is eroded. By measuring the height of the mounds and the age of the trees from tree-ring counts they estimated a soil lowering of about 10 mm/year (RAPP *et al* 1972). However, based on later research in Botswana, BIOT (1990) calculated that the rate of denudation as calculated by this method is 10 to 15 times greater than estimates by other methods. He offers the alternative explanation that the tree mounds can be explained by a difference in bulk density between soil in the mounds and the surrounding flat soil. He concluded that the mound results from a raising of the local surface rather than erosion of the surrounding surface.

Exposed tree roots may offer a valid indication of change when the reason is obvious, such as erosion in a streambed below a paint collar, but exposed tree roots offered as evidence of sheetwash, or of wind erosion in dry climates, should be treated with caution for Biot's hypothesis may also apply. Very long-term rates of erosion (over several centuries) were estimated from tree root exposure in Colorado (CARRARA and CARROLL 1979).

Clumps of grass elevated above the surrounding soil surface should also be treated with caution for the change may be the result of the grass trapping soil particles splashed from the surrounding soil. This was conclusively shown in Zimbabwe where erosion was measured from runoff plots under various tobacco/grass rotations. After a few years the tufts of weeping love grass (*Eragrostis curvula*) were found to be several centimetres higher than the soil surface between, although the measured soil loss from the plot was negligible. Some simple tests with splash boards showed that there was no net soil loss from the plot, but considerable translocation of soil within the plot. Clearly it is necessary to be certain whether changes in soil surface level are the result of erosion down or elevation upwards.

8.2.3 Profile measurements

To measure small changes in surface level along a cross section, such as an area with a number of parallel cattle tracks, a profile meter may be suitable. (The case of larger changes as in rills and gullies is discussed in the next section on volumetric measurements.) The requirements for a profile meter are to be able to set up a datum from which changes in level can be measured along a straight line and which can be re-established at the same points later to measure changes in level. Usually this takes the form of a horizontal bar with rods which can be lowered down to the soil surface, and is the same principle as that used to measure surface roughness in studies of tillage and tilth. Such a device to measure surface levels accurately on grazing land was developed by the author (HUDSON 1964). Metal pegs were set unobtrusively at ground level in concrete blocks at intervals of 2 m. A light aluminium girder could be fitted onto any two adjacent pegs and this gave a firm datum from which the level to the soil surface could be accurately measured at positions marked on the girder. Between readings the girder was removed so that there was no interference with cattle movements. Measurements were taken to the nearest millimetre, which allowed annual changes to be clearly recognized. Another simpler device is shown in figure 8.2. A similar device has been used by MCCOOL *et al* (1981). In this case a large number of pins are lowered to the soil surface at the same time and the profile is recorded by camera for later evaluation. Several other more sophisticated profile meters have been developed and details are given in *Further reading*.

8.2.4 Volumetric measurements

Estimates of soil loss based on three-dimensional measurements of volume can be used in different ways. For erosion from rills or roads we measure the length of the eroding section and changes in cross-sectional area. For gully erosion we usually want to know not only the volume lost, but also how much the gully is increasing, so we also have to measure changes in length as the gully cuts back. The other volumetric approach is

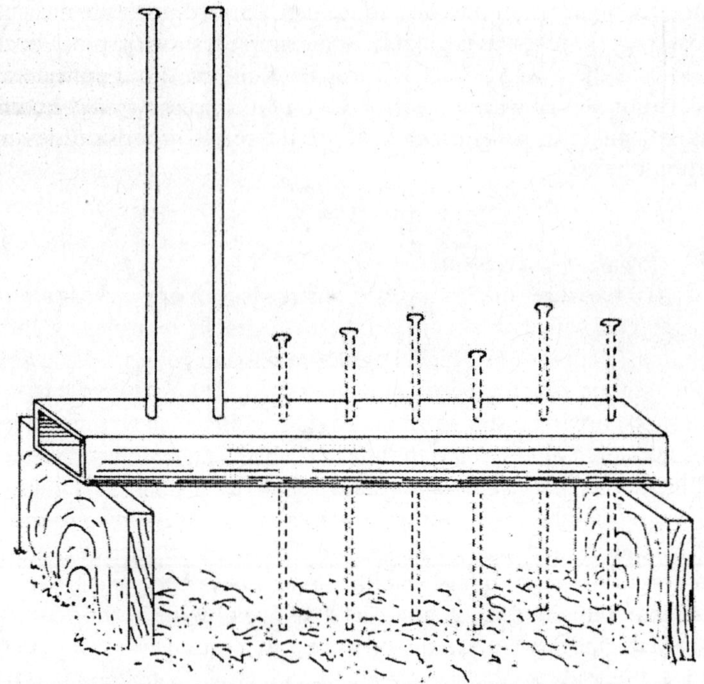

Figure 8.2 A simple profile meter for measuring changes in surface level (FAO)

to measure or estimate the volume deposited as an outwash fan (plate 8.1), or in a catchpit or reservoir (plate 8.2).

Rills and roads

Measuring the cross-section of all the rills in a sample area or along a sample transect is quick and easy, so the method is suitable for measuring change over short time periods, such as the change caused by a single heavy storm. The cross-section may be estimated from measurements of average width and depth if the shape is fairly uniform, or by summing the area of segments if the cross-section of the rill is irregular.

Gullies and streambanks

When the progress of gully erosion is being studied, measurements are needed both of the horizontal spread of the gully and vertical changes within the gully. To measure the surface area, and changes from cutting back or bank collapse, a rectangular grid of erosion pins is set out at an appropriate grid interval of perhaps 2 or 5 m as in figure 8.3. From

Plate 8.1 Volume of soil lost from a rill or gully compared with the volume of deposited soil (R. Evans)

Plate 8.2 Plastic-lined catchpits in Thailand (S. Sombat Panit)

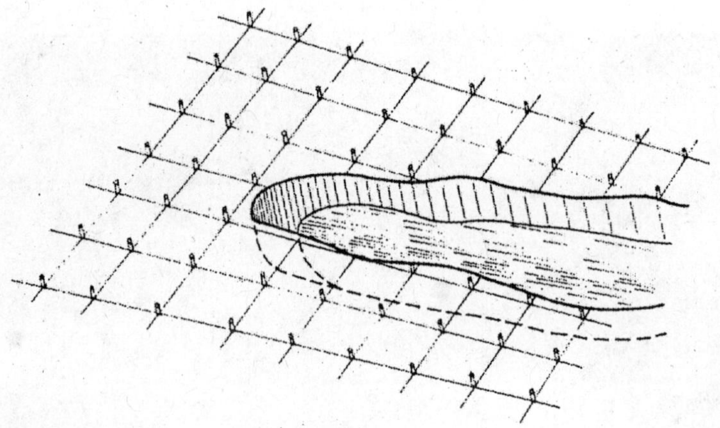

Figure 8.3 Setting out a network of erosion pins to measure gully erosion

measurements along the grid lines from the nearest pin to the gully edge, the surface area can be plotted on graph paper. The grid lines also serve as the transects for cross-sections across the gully. A string is stretched at ground level along a grid line with markers at fixed intervals of, for examples, 1 m. At each marker the depth is measured from the gully floor using a survey staff or a ranging rod, and the section can be plotted. The volume of soil lost from the gully is calculated as in figure 8.4, and subsequent measurements will quantify the changes.

Changes in a gully may be interpreted from the use of sequences of photographs. The position of the camera and the direction of the photograph must be carefully recorded. It is surprising how seldom 'before and after' photographs of gullies are lined up accurately. For studies of the long-term development of gullies, aerial photography can be a useful tool. An interesting example from Zimbabwe allowed the correlation of the changes in a gully with the changes in land use and vegetation in its catchment over a period of 40 years (KEECH 1989).

Catchpits

Surveys of sediment in reservoirs can be used to make quantitative estimates of erosion as discussed in HUDSON (1993), but simple catchpits may be used to demonstrate comparisons (plate 8.2). It is not possible to get a reliable estimate of the total soil movement unless the receiving reservoir is large enough to contain the whole flow and sediment load, but smaller pits which only catch an unknown proportion of the sediment can still be used to obtain comparative information. This was

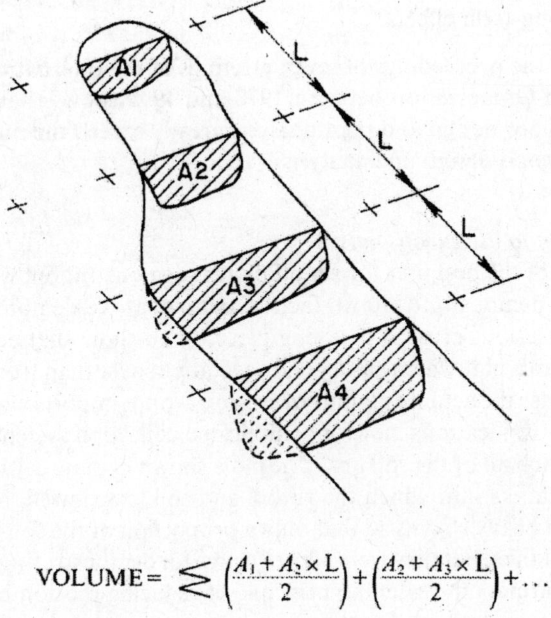

Figure 8.4 Calculation of cross sections in a gully

done successfully in the FAO project in Java previously referred to in connection with erosion pins (FAO 1976), where small catchpits were dug on two small parallel catchments, one of which was terraced and the other not. Previously sceptical farmers were convinced of the effectiveness of terracing when they saw that there was much less soil in the catchpit below the terraced plot than the untreated plot.

8.3 Experiments with field plots

It has to be said that there is today a large and growing number of field experiments using runoff plots, but the overall level of effectiveness is disappointingly low. The reasons are discussed in detail in FAO *Soils Bulletin 68* (HUDSON 1993) but the main weaknesses are:

- Assuming that results from small plots can be extrapolated to large areas.
- Inadequate replication.
- Inadequate randomization to eliminate bias.

- Assuming that results from one or two seasons are representative of long-term effects.

A study of the proceedings of seven meetings of the International Soil Conservation Organization between 1978 and 1992 shows that reports which show poor design and poor analysis greatly exceed the number of studies with good design and analysis.

When to use runoff plots

One of the best uses for runoff plots is demonstration, where the purpose is to demonstrate known facts. Examples are to demonstrate to farmers that serious erosion is taking place, or to show that erosion is much less from a plot which has a good vegetative cover than from a bare plot. In this case the actual amounts of erosion are not important, so there is no need for replications, nor for complicated collection systems which attempt to catch all of the soil lost. The plots shown in plate 8.1 had only simple brick tanks into which the runoff and soil loss flowed, and they overflowed in heavy storms so that only a proportion of the soil lost was caught in the tanks, but they were very effective for demonstrating to large numbers of farmers the essential principle of reducing erosion by better cover.

Another valid use is in comparative studies, for example to test or demonstrate, or get an approximate indication of the effect on runoff or erosion of a simple comparison such as with or without a surface mulch (plate 8.3), or the amount of runoff at the top and bottom of a slope.

A third possible use is to obtain data which are to be used to construct or to validate a model or equation to predict runoff or soil loss. The difficulties in collecting data of sufficient accuracy and reliability are great and so numerous that only large experimental programmes conducted at great expense over a long period of time can really meet this objective.

The problems associated with runoff plots are many and varied:

- Runoff plots are expensive, both in the initial construction and in their maintenance and operating cost.
- They use up a great deal of staff time at several different levels. There is much unskilled manual work in farming the plots and applying the treatments and emptying the tanks, but all of this has to be carefully supervised or things will go wrong.
- Easy access to the site is important, but very often the conditions which are to be investigated can only be found in remote areas. Complete

Plate 8.3 Micro-plots are useful for establishing the order of magnitude of the difference in runoff and erosion between simple treatments – in this case with and without a surface mulch. The stump in the centre of each plot is a newly planted tea bush

reliability is only achieved when professional staff can get to the site quickly at all times of day and night and in all weathers.

- Back-up facilities are needed. Laboratory facilities will be required for handling the samples, and technicians to repair electrical or mechanical equipment.

- Runoff plots have all the problems and difficulties of agronomic trials but in addition the much more difficult problems of collecting, catching, and recording the soil and water. There is a huge scope for faults and errors.

There are also constraints on what can be investigated on small runoff plots. Cultivation and other farm operations using tractors or oxen are difficult, and so are treatments which involve livestock, but increasing the size of the plot to allow realistic farm operations means dealing with large volumes of runoff and soil loss. Treatments which cannot be investigated on small plots include conservation measures which involve substantial earthmoving such as bench terraces or channel terraces. The hydrologic effects of channel terraces cannot be reproduced on small plots because the mechanics of runoff are completely different. Similarly the effect of bench terracing is totally artificial on a small plot because lateral movement of surface water is inhibited by the plot boundaries. The field scale loss of soil from land with bench terraces depends on the probability of the structure failing, and that is not something which can be studied on small plots. Measuring the effect of barriers set out on the contour also faces the problem of the unknown probability of failure. Examples are agroforestry practices such as hedgerows and alley cropping, trash lines, and contour furrows. An assessment of these on runoff and erosion can only be made on field-scale plots.

8.3.1 Types of runoff plots

Natural or simulated rainfall The cheapest and simplest method is to install the plot and then wait for rain, but the unpredictability of rain can make this frustrating. The alternative is to use artificially manufactured rain through the use of rainfall simulators as discussed in section 8.4. The main advantages of using a rainfall simulator are that it can speed up getting results, and the amount and type of rainfall can be controlled. But repetition of simulator runs on the same plot, while improving matters, is not a substitute for replication because it does not eliminate bias from soil variation. The disadvantage of simulators is that simulators for large plots are expensive to build and have a high labour requirement to operate. Simpler and cheaper simulators are usually restricted to small plots of a few square metres and these do not reproduce real conditions of surface flow.

Bounded or unbounded Most plots have boundaries which define the area from which the runoff and soil are being collected, but there are some cases where it is appropriate to use unbounded plots using what are usually referred to as *Gerlach troughs* after their inventor. They consist of a small collecting gutter which is let into the soil surface and connected to a small collecting container on the downstream side. There are various degrees of sophistication in the construction of the gutters and containers but complicated construction is not justified because what is required is a large number of replications to overcome the variation which arises from the fact that without any boundaries to direct or limit runoff into the collecting gutter, the amount collected depends on the chance occurrence of minor depressions or rills.

8.3.2 *Size of plots*

The size of plots must be related to the purpose of the trial.

- Microplots of one or two square metres may be appropriate if the objective is a simple comparison of two treatments where the effect of those treatments is unlikely to be influenced by scale. An example is illustrated in plate 8.3 where the objective was to demonstrate and obtain an approximate figure for the difference in surface runoff when a grass mulch was applied to newly planted tea bushes.

- Small-scale plots usually of about 100 m^2 are most commonly used for trials of cropping practices, cover effects, rotations, and any other practice which can be applied to small plots in the same way as it would be on a field scale, and where the effect can be expected to be unaffected by plot size. The original size and shape for this type of plot adopted in the United States was extremely arbitrary – 6 feet wide seemed to be a suitable width, and an area of 100th of an acre would be a convenient size for subsequent calculations, and this corresponded to a length of 72.6 feet which was adopted. There is some justification for following a well-established practice so that direct comparisons may be made, but there is no need to follow these precise measurements in metric units. In fact, plots only 6 feet wide are liable to have a significant border effect, and a more sensible size in metric units would be 5 metres wide and 20 metres long.

- Field plots of about one hectare are appropriate for assessing treatments which cannot be applied realistically to small plots. It is possible to apply cultivation and other farm operations through the use of removable plot boundaries, as shown in plate 8.4, but this requires large areas between the plots for turning oxen or tractors, and also extreme care in the replacement of the plot boundaries after each operation. A major source of error in runoff is leakages across plot boundaries so it is unwise to increase this risk by frequent alteration to the boundaries. Plots of the order of one hectare or more are necessary to assess any form of terracing, and also to assess the effect of grazing or livestock management.

8.3.3 *Measuring equipment*

In the case of small plots, all the runoff is led into a single collecting tank where it is stored until it can be measured, sampled, and recorded. For larger plots, or when large amounts of runoff are expected, it is impractical to store the whole of the runoff, and some device is used to divide it accurately so that a known fraction can be separated off and stored.

Plate 8.4 Removable plot boundaries are necessary if tractor tillage is part of the plot treatment. In this case planks of asbestos-cement are set in the ground

There will always be floating organic material in the runoff, and this must be caught on screens if any type of divisor or sampler is used. Examples can be seen in plate 8.5. A device widely used in the United States for many years is the GEIB Divisor, which consists of a number of equal rectangular slots. The water passing through the central slot is collected and stored, while that through the other slots runs to waste (plate 8.6). This type requires a high degree of accuracy in its manufacture and so a number of simpler alternatives have been developed. These include a series of V-notches (plate 8.7), or a vertical row of holes drilled in a steel plate (plate 8.5), or a series of pipes built into the wall of the tank as in plates 8.8 and 8.9. Any of these can be constructed to take a sample of between one fifth and one twentieth of the flow. Another approach is to split the flow successively by dividing it in half and then dividing again by half as many times as required, and an example is shown in plates 8.10 and 8.11.

There are several types of divisor which involve moving parts. These are

Plate 8.5 Screens trap floating debris and the divisor takes a 1/15th sample by collecting the flow through one of 15 vertical rows of holes drilled in a stainless steel plate

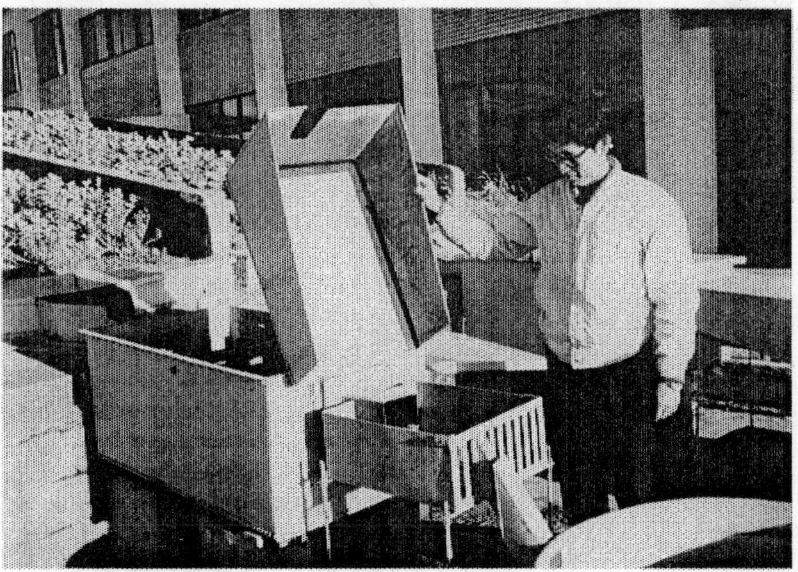

Plate 8.6 A multi-slot divisor in Korea

Plate 8.7 A divisor which takes a 1/10th sample by collecting the flow from one of ten V-notches. An HS flume (right back) measures the runoff into the tanks

only suitable where there will be constant supervision and maintenance, because there is a high risk of such divisors becoming jammed or choked by debris, or subject to mechanical failure. One of the best known is the *Coshocton wheel* sampler shown in plate 8.12. This is installed under the discharge from a flume, and the force of water turns a rotating slot sampler mounted on a vertical axis. Any recording device which depends upon moving parts or electrical supply is only suitable for use on experimental stations where there will be constant supervision, particularly during the extreme event, which is when mechanical and electrical devices fail, and usually occurs in the middle of the night. It should be pointed out that there have been great improvements in instrumentation in recent years. In most circumstances nowadays, electrical or mechanical recorders for water level would be replaced by pressure transducers feeding solid-state memory packs. The choice of recording equipment should depend on the special circumstances of each project or experiment.

Measuring rate of runoff is fairly straightforward. There is a wide range

Plate 8.8 A multi-pipe divisor in Indonesia (A. Mitchell)

of standard flumes available – standard flumes meaning those which, when built and installed according to specified conditions, do not need to be individually calibrated because the flow rate can be read directly from tables or charts if the depth of flow is measured. The most commonly used flume for small plots is the H flume which was designed specifically for this purpose by the United States of America Soil Conservation Service (USDA 1979). A range of flumes of different size has capacity from 0.0028 m^3/s to 3.08 m^3/s and plate 8.13 shows one of the larger type used to record the runoff from a plot of two hectares. The shape of the notch gives the desirable combination of sensitivity at low flows and large capacity at high flows. The backward slope of the notch makes the flume unlikely to be blocked by floating debris, and the floor is non-silting. A free fall from the

Plate 8.9 A multi-pipe divisor in Nigeria (R. Lal)

Plate 8.10 A simple flow-splitter in Thailand

Plate 8.11 Splitting the flow in sequence on plots in Thailand

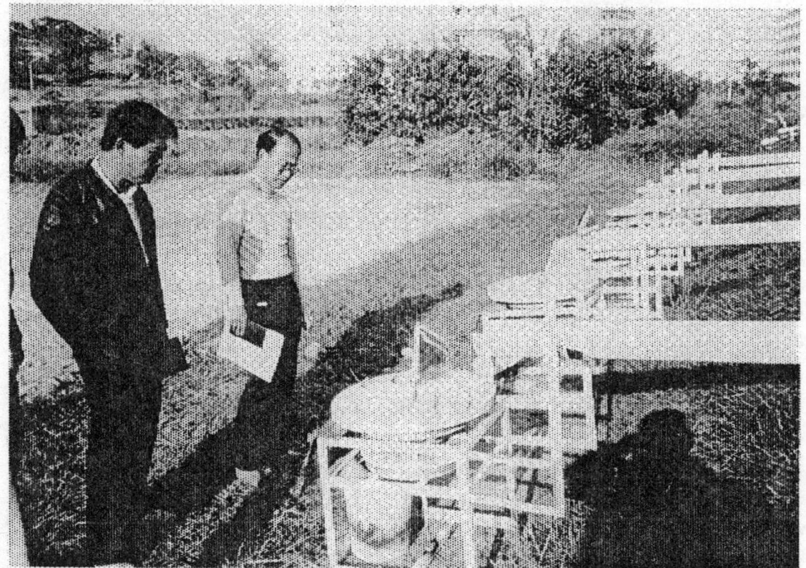

Plate 8.12 The Coshocton sampling wheel on runoff plots in Taiwan

Plate 8.13 An H-type flume recording runoff from the maize plot behind, which is 2 ha

outlet side of the flume is preferable, but when the available head is limited, the flume can be installed to flow with partial submergence, and calibration corrections for this are also provided up to 85% submergence. The rate of flow through the flume is a function of the depth of flow, and this is measured in a well-chamber connected to the wall of the flume. Spot measurements of depth can be made at a gauge plate or, for continuous recording, a liquid-level recorder is installed, formerly float-operated but nowadays more likely to be a pressure transducer linked to a solid-state memory.

Another well-tested flow-measuring device is the Parshall flume, and construction details and rating tables are available for flumes catering for a wide range of flows (PARSHALL 1950). This flume, like the H flume, can be operated either with free flow or with partial submergence, but in the latter case measurements of the depth of flow are required at two points in the flume (plate 8.14).

Having collected in the storage tanks a mixture of the soil and water lost

Plate 8.14 A Parshall flume with water level recorders

from the plot (or a known proportion of it) the mixture must then be separated so the soil and water may each be measured. One method is to add a flocculant, which causes the suspended matter to settle, so the clear supernatent liquid can be drawn off. This leaves a thick sludge from which a sample may be taken (JACKSON 1964). Very effective flocculating chemicals are now commercially available. Another method avoids the time-consuming drying and weighing in the laboratory by weighing a fixed volume of the sludge and comparing this with the weight of an equal volume of water (ELWELL 1976). If divisors are used, it is necessary to calculate the weight of soil in each of the tanks separately because the concentration of sediment will be greatest in the first tank where the heavy particles settle out, and least in the final tank. If it is likely that the relationship between runoff and soil loss will be similar for the treatments being compared, it may suffice to measure runoff only, which is much easier than measuring soil loss.

8.4 Rainfall simulators

8.4.1 The advantages and objectives

Field plot experiments depend upon natural rainfall which is always unpredictable and frequently perverse. For many years research workers have sought to be independent by using a man-made simulation of rainfall. This has two advantages, both very important. The speed of research is greatly accelerated since the results are no longer dependent upon waiting for the right kind of rain to come at the right time, and also the efficiency of the research is increased by control of one of the most important variables, rainfall. It is no longer necessary to interpolate or extrapolate from storm to storm – the same storm can be created over and over until the results have been tested and confirmed. Another minor advantage is that it is usually quicker and simpler to set up a simulator over existing cropping treatments than to establish the treatments on runoff plots.

The disadvantages are all related to scale. It is cheap and simple to use a small simulator which rains onto a test plot of only a few square metres, but simulators to cover field plots are large, expensive, and cumbersome. Measurements of runoff and erosion from simulator tests on small plots cannot be extrapolated to field conditions. They are best restricted to comparisons, such as which of three cropping treatments suffer least erosion under the specific conditions of the simulator test, or the comparison of relative values of the erodibility of different soil types. Simulators are likely to be affected by wind, but having to erect wind shields undermines the advantages of simplicity.

Simulators can be useful tools, for some purposes but quite unsuitable for others, and the objectives will dictate the most appropriate type of simulator. For example, in studies of infiltration and runoff it is not necessary for the simulated rainfall to have precisely the same characteristics as natural rain. In other studies it may be important that the erosion processes are not distorted by the simulated rain being different from natural rain (MEYER 1965). The required size of the test plots may dictate the type of simulator, *eg* small plots may be suitable for studies of relative erodibility, but larger plots would be required for measuring rill erosion.

Simulators are appropriate for studies of the relative protection afforded by different plant densities and at different times during the growing season, or relative erodibility, soil infiltration characteristics, and erosion and runoff from up-and-down-slope row crops.

Simulators are not suitable for studies of crops grown on a contour, because the plot borders interfere with the normal water flow. Nor are they suitable for the comparison of treatments which have only minor differences, because under field conditions, experiments with rainfall simulators will suffer from large uncontrollable experimental variation.

8.4.2 Construction and operation

There are few commercial suppliers of rainfall simulators, so it is usual for research workers to build their own. However, there is a large amount of literature reporting the building and testing of rainfall

Plate 8.15 A laboratory simulator where the intensity can be controlled by the height of the overflow pipe, designed at the University of Ghent, used here at the Central University of Venezuela, Maracay

simulators so it is usually practical to copy a previous design rather than to start from the beginning. In particular, a huge variety of commercially available spraying nozzles has been tested. The two basic principles of making raindrops is from nozzles under a low static head of water or pumped through nozzles under pressure.

Low pressure droppers

Many rainfall simulators have used the principle of drops forming and dropping from the tip of tubes connected to a water supply. The size of the drop is related to the size of the tube. Metal, glass, or plastic tubing has been used, or hypodermic needles, which are manufactured to a high degree of accuracy. An array of tubes of different sizes may be used to produce rain with different drop sizes. Constant pressure must be maintained in the water supply to the nozzles, either by using a constant-head overflow as in plate 8.15, or by the principle of the Marriotte bottle (KAMPORST 1988). The advantages of this method are that the size of the drops and their fall velocity are constant, and the distribution of rainfall across the test plot is uniform and can be achieved with low water

Plate 8.16 A trailer-mounted low pressure dropper simulator in Venezuela

pressures. Sometimes a random distribution of rainfall onto the test plot is achieved by a mechanism to oscillate the drop-former or to rotate it.

The disadvantages are that unless the device is raised up very high, the drops strike the test plot at a velocity lower than the terminal velocity of falling rain, and therefore the values of kinetic energy are low. A large drop of 5 mm diameter needs a height of fall of about 12 metres to reach terminal velocity and this is difficult to achieve in field conditions. To some extent this can be compensated by using larger drops than occur in natural rainfall. Another disadvantage is that the size of the test plot is limited by the practicalities of constructing a very large drop-forming tank. A simulator using this approach and mounted on a small trailer has been successfully used for many years in Venezuela (plates 8.16 and 8.17).

Spraying simulators

Many types of spraying nozzle are commercially available, some designed for other purposes and some designed especially for rainfall simulators. A major difficulty is that if the spray is to include drops of the largest size which occur in natural rain, then the nozzle opening has to be large – about 3 mm diameter. But even with low water pressures the intensity of rain produced from nozzles of this size is higher than natural rain (ELWALL and MAKWANYA 1980). It is therefore necessary to have some kind of interruption of the spray to reduce the intensity to that of natural rain. In Meyer's *Rainulator* two methods were used (figure 8.5). The spray

Plate 8.17 Preparing plots for the Venezuela simulator

nozzles were mounted on an overhead carriage which traversed backwards and forwards across the plot, and also the flow of water to the nozzles was switched on and off by solenoid valve (MEYER and MCCUNE 1958). This simulator and its derivatives are very efficient, but because they were designed for operation on large plots they are complicated and expensive. Most subsequent developments have therefore been concerned with designing simpler or smaller machines. One such variation was designed by DUNNE *et al* (1980) for field use in Kenya. A trolley carrying a spray nozzle is pulled backwards and forwards along an overhead track by two operators pulling alternately on ropes. Another method is to mount a row of downward-spraying nozzles on a delivery pipe which oscillates through an arc of about 120° as in the device used by MEYER and HARMON (1979) and shown in figure 8.5.

Another approach is based on a commercial rotating-boom irrigation machine shown in figure 8.5 and plate 8.18. Each boom carried the water supply to a number of nozzles on each boom which rotate slowly, powered by a water turbine. The machine is set up between two test plots so that rain can be applied simultaneously to both plots. Plot lengths up to 15 m can be rained on by one machine, or for longer plots two machines can be used (SWANSON 1965). A novel approach to the problem of reducing the

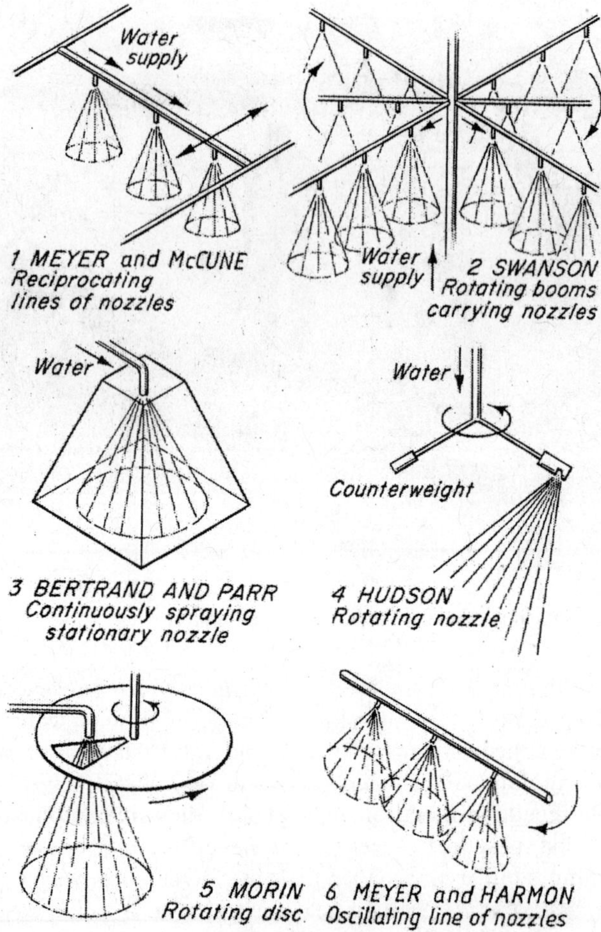

Figure 8.5 The working principle of some rainfall simulators

intensity was introduced by Morin and his co-workers (MORIN *et al* 1967). In their simulator a fixed nozzle sprays continuously, but the soil is intermittently shielded from the spray. The nozzle is directed vertically downwards, and just below it is a metal disc which rotates in the horizontal plane. A radial slot is cut in the disc, and each time this passes under the nozzle a short burst of rain passes through to the plot below. The proportion of the spray which passes is determined by the angle of the slot. This design allows the use of large nozzles which give the right drop size distribution and kinetic energy but which produce excessive

Plate 8.18 The rotating-boom simulator, made from a modified irrigation machine by SWANSON (1965)

intensities when spraying continuously. There are many modifications and variations on Morin's basic design. In some the speed of rotation of the disc or the width of the slot can be varied during operation, so that the instrument can be programmed to reproduce storms with varying intensity. Another variation is to cut several slots in the disc so that the frequency of the bursts of rain can be increased while keeping the speed of revolution low.

An alternative way of achieving intermittent application of rain from large nozzles is to use the reaction of a nozzle to rotate an upside-down irrigation sprinkler. An early device of this type is shown in figure 8.5 and plate 8.19. This was designed to be light and portable so that it could be taken to remote areas with poor road access. After the machine is set up it can be positioned in turn over each of the six hexagonal plots which are arranged in circular pattern around the supporting mast. Six replications of each test can thus be made without moving the machine. A later model suitable for larger plots was developed at Silsoe College (plate 8.20). A major advantage is that this type of machine can be assembled from off-the-shelf components (plate 8.21).

Most modern simulators are based on intermittent application of rain, but continuously spraying simulators have been built which achieve the required rainfall characteristics by injecting air into the water supply to

Plate 8.19 A small rotating-nozzle simulator. The inverted Y rotates under the reaction of the jets. Tests can be carried out on each of the hexagonal plots, giving six replications at each test site

nozzles. The air/water mixture and the pressure can be varied to control the intensity and drop size distribution (SHELTON *et al* 1985, HINKLE 1990).

Operation

Runoff plots used with rainfall simulators are the same as miniature runoff plots discussed earlier, and the same considerations apply to plot boundaries, a collecting trough leading the runoff and sediment to containers, and recording the volume of runoff and weight of soil.

Plate 8.20 The Silsoe College rainfall simulator (R. J. Rickson)

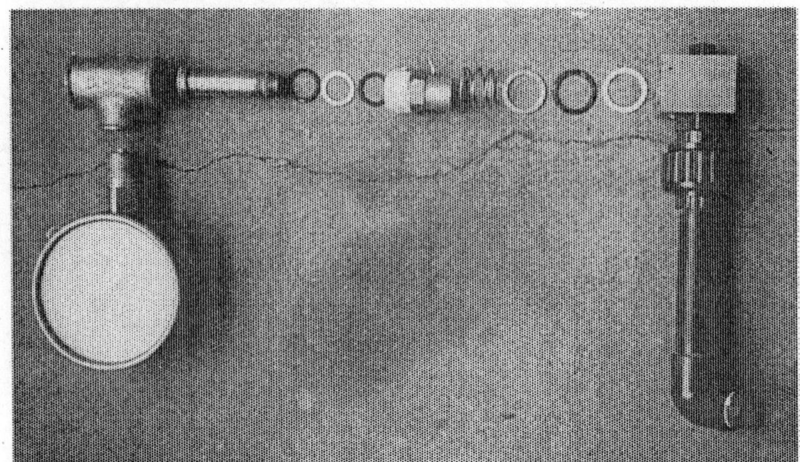

Plate 8.21 The components of the rotating nozzle simulator developed at Silsoe College (R. J. Rickson)

Much time and effort is required for setting up some of the larger machines and some simulators are designed to reduce this. The Australian simulator in plate 8.22 is made of lightweight materials, and can be picked up in one piece by a mobile crane and swung onto a new plot in minutes. The rotating boom in plate 8.18 covers two plots from one position, and the rotating nozzle in plate 8.19 allows six replicated plots for each setting up.

Tests have shown that results from simulator trials are considerably affected by the initial soil moisture of the test plot, and to reduce this undesirable variable it is usual to specify some standardized pre-wetting treatment. One method is to saturate the soil 24 hours before the tests so that the soil is approximately at field capacity. An alternative is to apply 25 mm of rain at 100 mm/h four hours before each test. The amount of simulated rainfall during a test must be measured. One method is to cover the whole plot with a collecting sheet or pan of the same size and measure the rain caught during a fixed period. This is done before and after the test to make sure there is no change during the test. Alternatively, measurements can be made during the test either by small rain gauges installed on the test plot, or by collecting channels across the plot as in plate 8.18.

References

BIOT, Y. 1990 How long can high stocking densities be sustained? Paper to *Technical Meeting on Savannah Development and Pasture*

Plate 8.22 A large lightweight Australian simulator which can be picked up by a crane in one piece and moved to an adjoining plot

Production, Commonwealth Secretariat and ODI, Woburn, November 1990

CARRARA, P. E. and T. R. CARROLL 1979 The determination of erosion rates from exposed tree roots in the Piceance Basin, Colorado, *Earth Surface Processes*, 4, 307–317

DUNNE, T., W. E. DIETRICH and M. J. BRUNENGO 1980 Simple portable equipment for erosion under artificial rainfall, *Journal of Agricultural Engineering Research*, 25, 161–8

ELWELL, H. A. 1976 A rapid method for estimating the dry mass of soil from erosion research plots, *Research Bulletin 20*, Department of Conservation and Extension, Ministry of Agriculture, Harare, Zimbabwe

ELWELL, H. A. and H. MAKWANYA 1980 Design and calibration of a rainfall simulator nozzle assembly for laboratory and field studies, *Research Bulletin 25*, Department of Agricultural Technical and Extension Services, Zimbabwe

FAO 1976 *Upper Solo watershed management and upland development project, Indonesia*, Project INS/72/006, Termination Field Document 6, Soil Conservation, by B. C. John and W. van der Goot, FAO, Rome

HAIGH, M. J. 1977 The use of erosion pins in the study of slope evolution, in *Shorter Technical Methods (II), Technical Bulletin 18*, British Geomorphological Research Group, Geo Books, Norwich, Norfolk

HINKLE, S. E. 1990 The Akron USDA-ARS portable boom-mounted rainfall simulator, *Transactions of the Association of Agricultural Engineers*, 33, 3, 818–820

HUDSON, N. W. 1964 Field measurements of accelerated soil erosion in localized areas, *Rhodesia Agricultural Journal*, 31, 3, 46–48

HUDSON, N. W. 1993 Field Measurement of Soil Erosion and Runoff, *Soils Bulletin 68*, FAO, Rome

JACKSON, D. C. 1964 Sludge sampling techniques for soil erosion research, *Research Bulletin 12*, Department of Conservation and Extension, Salisbury, Zimbabwe

KAMPHORST, A. 1988 A small rainfall simulator for the determination of soil erodibility, in *Land Conservation for Future Generations*, edited by S. Rimwanich, Department of Land Development, Ministry of Agriculture, Thailand

KEECH, M. A. 1992 The photogrammetrical evaluation of areal and volumetric change in a gully in Zimbabwe, chapter 5, in *Erosion, Conservation, and Small-Scale Farming*, edited by K. Tato and H. Hurni, Geographica Bernensia, Switzerland

MCCOOL, D. K., M. G. DOSSETT and S. J. YECHA 1981 A portable rill meter for field measurement of soil loss, in *Erosion and Sediment Transport Symposium*, IAHS Publication 133

MCFARLANE, D., N. DELROY and S. VREESWYK 1991 Water erosion of potato land in Western Australia, *Australian Journal of Soil and Water Conservation*, 4, 1, 33–40

MEYER, L. D. 1965 Simulation of rainfall for erosion control research, *Transactions of the American Society of Agricultural Engineers*, 8, 1, 63–65

MEYER, L. D. and D. L. MCCUNE 1958 Rainfall simulator for runoff plots, *Agricultural Engineering*, 39, 10, 644–648

MEYER, L. D. and W. C. HARMON 1979 Multiple intensity rainfall simulator for erosion research on row sideslopes, *Transactions of the American Society of Agricultural Engineers*, 22, 1, 100–104

MORIN, J. D., D. GOLDBERG and I. SEGINER 1967 Rainfall simulator with a rotating disc, *Transactions of the American Society of Agricultural Engineers*, 10, 1, 74–77, 79

PARSHALL, R. L. 1950 Measuring water in irrigation channels with Parshall flumes and small weirs, *Circular 843*, United States Department of Agriculture, Soil Conservation Service

RAPP, A., D. H. MURRAY-RUST, C. CHRISTIANSSON and L. BERRY
1972 Soil erosion and sedimentation in four catchments near Dodoma, Tanzania, *Geografiska Annaler*, Series A, 54A, 255–318

SHELTON, C. H., R. D. VON BERNUTH and P. RAJBHANDARI 1985 A continuous application rainfall simulator, *Transactions of the American Society of Agricultural Engineers* 28, 4, 1115–1119

SWANSON, N. P. 1965 Rotating-boom rainfall simulator, *Transactions of the American Society of Agricultural Engineers*, 8, 1, 71–72

UNITED STATES DEPARTMENT OF AGRICULTURE 1979 Field Manual for Research in Agricultural Hydrology, *Agricultural Handbook 224*, USDA, Washington DC

9

Land use and soil conservation

9.1 Planning land use

Modern techniques such as mechanization, better crop varieties, and the scientific use of fertilizers can transform agriculture, but before they can be effective the use of the land has to be right. No techniques will make it possible to grow a good crop if the soil conditions are unsuitable for that crop, and no conservation works can prevent erosion when the basic cause is trying to grow crops on land which is really unsuitable for arable farming (plate 9.1). The object of land use planning is to determine the characteristics of land, its possibilities, and its problems, and try to match these to a form of land use which will be sustainable and economically productive.

9.1.1 Some planning principles

An often quoted definition is that 'planning is the conscious process of selecting and developing the best course of action to accomplish an objective'. In the case of Land Use Planning the objective is the efficient intensive use of land resources.

It is difficult to imagine any field of human activity where some form of planning is not essential to achieve the objective. Consider going on holiday. Certainly it is possible to come home from work one evening, throw some clothes into a suitcase, go to the railway station or airport and buy a ticket to the first holiday resort which comes to mind. Some fortunate people may indeed enjoy so unplanned a holiday but for most of us it would be disappointing to find oneself in a mountain holiday resort without climbing boots, or on a sunny beach with only winter clothing in the suit case. No, one is more likely to achieve the objective of a happy holiday by a planned approach and this means going through a logical sequence of steps, i.e.:

Plate 9.1 On these steep slopes in Bhutan, the erosion has been partly controlled by terracing, but the basic problem is that the land is too steep for arable farming

- Collecting all the relevant facts.
- Analyzing the facts.
- Making decisions.
- Carrying out the decisions.
- Assessing the results.

All five of these steps are essential both for our simple case of taking a holiday, and for planning the optimum use of thousands of hectares of land. Collecting the necessary facts for our holiday means finding out how much leave we have due, how much money we can afford to spend, which holiday resorts are suitable, and a host of equally obvious items. Next we must analyse these data, sort them into groups of related items such as the cost per day, the number of days, and the total cost. This analysis will allow us to set up options, such as 10 days in an expensive hotel at one resort, or 20 days in a cheaper hotel at another resort. Now we are in a position to take the third step – making the decision. In doing this we have to balance what is theoretically desirable, and what is practically possible, and we have to reject unacceptable or unworkable solutions. Arriving at the best decision is unlikely without the previous steps of collecting the data, and sorting them out. Next we carry out the decision by going on holiday, and finally we evaluate it. We consider: 'Was it worth it?' and 'Would I go there again?', or perhaps we might only learn from experience how to do it better next time.

In agricultural development also there is the unplanned approach of rushing in and starting schemes without surveys or pilot trials and no data collection, or there is a properly planned approach. In the planned approach the first step is collecting all the relevant information about the land and climate. Next the data are analysed into groups and classes which provide a systematic filing system for the data. When the land has been investigated and classified, the options can be weighed up and a decision made. Then comes the fourth step of carrying out the plan, and finally assessing the result to see if the objective has in fact been achieved. Governments in particular seem liable to forget that carrying out the plan does not automatically mean achieving the objective. The last stage of evaluation is as vital as any other, for there may have been some unknown factor which has prevented the plan from working.

Of course, this is not the whole story of land use planning. There are economic, political, and social aspects which must be considered, but the two points we are making here are that first, there must always be a logical sequence of steps in the planning process, and second, there has to be some systematic method for converting the basic data into usable form. In the rest of this chapter we will look at some of the more common methods for doing this.

9.1.2 *Planning the use and development of natural resources*

Planning the sound use of land takes place at different levels, each of which has a separate set of objectives and its own time-scale. We will define these, and apply them to an example.

Setting a policy is defining the long-term objective – stating where you want to get to in some defined period, perhaps 10 or 20 years. Policies are determined by national considerations and so have to be set at national levels.

The strategy is deciding the general direction or route you intend to take towards the policy goal, for example going over a mountain, or around it, or tunnelling through it. The conditions influencing the selection of strategy will vary within the country according to different physical and ecological conditions. So the strategy will be more specific and perhaps apply only to a province or district. The time-scale will be shorter, perhaps a five-year rolling plan.

The tactics are the details of how the strategy is to be implemented. How you intend to build the road, and where a bridge is required. The tactics need to be more flexible, so are planned over a short timespan, perhaps two or three years.

Nowadays it is increasingly recognized that people participation is essential for the implementation of land improvement plans, and should be incorporated as early as possible in the formulation of plans, and this will be discussed later in this chapter and in chapter 14. However, we need more experience before being able to suggest how this can be achieved at all levels. Governments are not able to contribute to village level discussions on land use and management, and village councils are not able to contribute to government consideration of national politics and economics, nor to international issues such as trade and technical assistance.

We might look at Lesotho as a case study of land use planning. Lesotho is fortunate in having a detailed inventory of its natural resources, partly because of its small size and partly because of its colonial history. There is a good database on the physical facts such as geology, soils, climate, land systems, and the ecological factors such as present land use and recommended land use. To create a national policy for agricultural development it is therefore only necessary to combine the known capacity of these resources with the social, political, and economic objectives of the government. These are unique to Lesotho. For one thing it is totally surrounded by the Republic of South Africa, and also there is a large exodus of labour to the mines and industries of South Africa. This raises some special issues, such as whether the country should aim for self sufficiency in cereals, which is politically desirable but a drain on the national economy. Another special issue is the need for the forward planning of an agricultural industry to absorb the extra labour if at some time it is not employable in South Africa. A national policy has to face these issues, and they can only be decided by national government.

As in most countries, there are wide variations in agricultural potential, from high mountains suitable only for livestock production to lowlands with arable potential, some areas with irrigation capacity, some with good soils, some with poor soils. To cater for these variations, different strategic plans are required within the framework of the national policy. In Lesotho the place for strategic planning is the District Agricultural Development Plan (DADP) on a rolling five-year time-base, prepared with involvement of all the government departments and institutions operating at District level.

At the next level down is the tactical plan which in Lesotho may be a Ward ADP, a Community ADP, or a Village ADP. Flexibility is necessary because the size of wards and villages varies.

In this scenario the main input of the ideas, needs, and aspirations of the people are thoroughly discussed at community level and fed into the community plan. The interface with Government policy and strategic planning is when the District Development Plan is formulated by the Government departments in direct discussion with the elected community leaders.

9.1.3 Natural resource inventories

The need for an inventory The starting point for any natural resource policy has to be an inventory of the resources. It is not practical to start planning a steel industry without studies of how much iron ore and coal are available, and similarly a policy for agricultural development must be based on knowledge of the resources available. Regrettably, national agricultural policies tend to start with political goals, such as to be self-sufficient in a particular crop by a given date, without first assessing whether the resources of climate, soil, and labour allow this to be a realistic target.

On the other hand, an inventory does nothing on its own; it only becomes useful when it is used to help make decisions. Too many past inventories of natural resources have been just a collection of facts, and the end product was only a pretty coloured map to hang on the walls of government offices.

Another constraint is that people are one of the most important resources, and while it is possible to map or tabulate a population density or growth rate it is not so easy to record important information such as what are the reasons for the present farming systems? what are the land users' objectives? and what are the social or economic constraints that could hinder agricultural development such as prices, market outlets, or land tenure?

New techniques for data collection The compilation of a database of natural resources used to be long and expensive, and beyond the capacity of some developing countries, but a number of techniques now make it much simpler.

- Some of the old-style field surveys can be done more quickly and more efficiently by the use of remote sensing. For example, soil survey is greatly aided by the interpretation of aerial photography. Geology and topography may be assisted by the use of satellite imagery, and by airborne geophysical magnetic surveys (STREET and DUNCAN 1992). The collection of meteorological or hydrological data in remote areas can be improved by the use of telemetry and automatic recording.
- Data collection has been helped by improved sampling techniques and more sophisticated statistical interpretation of data from samples.
- The handling and processing of data is much better and quicker as new computer developments appear, and the presentation has also been greatly improved by the development of display equipment and printers.
- The ability to digitize information which is in map form means that maps can be combined in the computer without the laborious use of physical overlays.
- Equally important is the ability to combine different kinds of database such as maps, tabulated data, and remote sensing imagery.

Geographical Information Systems (GIS) The use of GIS is growing rapidly. This allows the computerized compilation of a single combined database which previously was contained in different forms, such as geological surveys, soil surveys, meteorological and hydrological data. Digitizing this information for computer storage not only improves the storage and retrieval but also allows the different sources of data to be integrated. This technique was first used on a grand scale by FAO for the combination of the data of the World Soil Map with climatological data to provide agro-ecological classifications and prediction of crop production and carrying capacity (FAO 1978).

There have been many recent developments and applications which enhance a GIS based on Landsat imagery by combining additional information into it. For example ZHOU (1989) shows how adding topographic data allows the mapping of classifications useful in land use planning such as slope classes, or by adding data on soil survey and climate enhances the ability to map crop production. Another Australian application combines detailed soil data with the GIS and the output can be in the form of either maps or printed text reports (ABRAHAM 1992). In New Zealand field survey data is added to GIS to allow computer-generated maps of Land Capability Classes for farm-scale planning (LIU *et al* 1992) and also in New South Wales, Australia (THOMPSON *et al* 1992).

Socio-economic surveys In current thinking about people participation in agricultural development, it is equally important to collate socio-economic data, and here too the methods and techniques have greatly improved in recent years. The former census-type data collection using pre-set questionnaires with pressure for the questionnaire to be completed, was never very efficient and more likely to alienate the people than win their support for development plans. Today's style is interactive or participatory surveys, and some of the approaches are briefly described in section 9.2.

9.1.4 Land classification and evaluation

There are many kinds of classification systems. Usually the purpose is to act as a kind of filing system so that collected data can be organized into packages or groups which can then be used for planning purposes. Some classification systems are purely *descriptive*, for example a soil classification which describes the properties of soils and says that soil type *A* has the properties *a*, *b*, *c*, and soil type *B* has the properties *d*, *e*, and *f*. But this is not helpful if the soil we are studying has the properties *a*, *b*, and *f*. For planning what is required is a *prescriptive* system, that is one which can synthesize rather than analyze. It enables us to look at the actual situation, and working from there says, 'because of the existing properties and conditions it must be class *X*, and therefore the possible land uses are *P*, *Q*, or *R*'.

Existing systems are variously known as *land classification systems*, *land suitability systems*, *land capability systems*, or *land evaluation systems*. The names are less important than the purpose, which is not always clear from the title, for example suitability by itself is meaningless until we know what it is suitable for. A classification of suitability for growing cotton will be different from a classification of suitability for growing tea. Land capability is also an imprecise label until defined. Its most frequent use is in the USDA-SCS system where the purpose is to assess the most intensive land use which can be sustained without soil degradation, and so has a built-in bias towards arable farming.

Another variable is whether the system considers only the existing situation, or whether the classification could be changed, for example by rectifying a nutrient deficiency through fertilizers, or excessive wetness by drainage.

The purpose of the classification system should be linked to the planning scale, whether national, regional, or farm scale. The classification of agro-ecological zones used for national planning in Zimbabwe is not unlike the system used in Kenya to define areas of high potential where intensification of agriculture is practical. On the other hand a classification designed to assist planning at the farm scale is much more detailed.

The inputs for different systems also vary. Simple systems such as the *World Life Zone System of Ecological Classification* may be based only on annual values of precipitation and temperature (TOSI 1964), and interestingly this is also the basis of the traditional and indigenous classification in Ethiopia (HURNI 1986). The national systems of agro-ecological zones in Zimbabwe and of potential in Kenya combine climate with soil data. For farm planning a broad assessment of climate will be sufficient for a whole region, but detailed soil data are required based on soil mapping units. Other inputs which may be included are the potential for productivity, or the constraints on it. If the purpose of the classification is to assist in land reallocation, as in the national schemes in India, Malawi, and Kenya, it may be appropriate to include land values, and socio-economic issues such as opportunities for off-farm employment, or labour availability.

9.1.5 An outline of some classification systems

Land Capability Classification Like other kinds of land classification, this system has a particular purpose, which is to record all the relevant data which will lead to a decision as to the combination of agricultural use and conservation measures which allow the most intensive agricultural use of the land without risk of soil degradation. Some of the key facts assessed are the depths of the soil and its texture, the land slope, and the past erosion. There are a number of Land Capability Classifications, for in every country and geographical region there are different factors which should be allowed for. The soils and climate will vary, and so will social customs, land tenure, economics – and all of these may affect the choice of the best land use. However, all these methods stem from that developed by the Soil Conservation Service in the United States (KLINGEBIEL and MONTGOMERY 1961). For a detailed account of the variations used in Zimbabwe, Israel, and the Philippines readers are referred to either of the two earlier editions of this book.

In the USDA system land is allocated into one of eight classes, of which the first four are suitable for cultivation and the other four unsuitable. The eight classes are ranked in order of increasing risk of erosion. It is commonly used together with the Universal Soil Loss Equation to estimate the soil loss under alternative cropping practices and conservation measures. A summary of each class follows.

Class I is land with little or no risk of erosion, and no limitations on use.

The soils are deep, productive, easily worked, and nearly level. Class I soils need only good farming practice to maintain soil fertility and soil structure.

Class II land carries some risk of damage so is subject to moderate limitations of use. It needs the same good farming practices as Class I and also simple conservation practices such as soil-conserving rotations or cut-off drains. The soils can be as productive as those of Class I, but are slightly downgraded because of shallower depth or steeper slopes.

Class III land has severe risks of damage so is subject to more serious restrictions of use. Soils are likely to be steeply sloping, less productive, and will require some form of terracing or cross-slope barriers. Cropping systems must concentrate on maintaining good plant cover. Rotations will be mainly grass and fodder crops with occasional cultivated row crops.

Class IV land is subject to very severe risk of erosion and needs full mechanical protection, and should be under semi-permanent cover crops most of the time, only occasionally used for a cereal crop, perhaps once in five or six years.

Class V land is not suitable for cultivation because of some limitation, such as severe wetness, which is not amenable to corrective treatment.

Class VI land is very steep, and use should be limited to grazing or forestry, and there may still be an erosion hazard under such use.

Class VII land is so steep, rough, or otherwise limited that even under grazing or forestry, careful management is needed to avoid serious erosion.

Class VIII land has such severe limitations that it is of negligible agricultural value, and should be used for wildlife, recreation, or water supplies.

Any class of land can be used for a use normally associated with a lower class, for example on a smallholding on the outskirts of a town it might be more appropriate to use Class I land for milk production from a high-yielding managed pasture than to use it for arable crops.

Applications and limitation There is an implicit assumption in Capability Classification that land should be used as intensively as possible without any degradation. This in turn implies that the farmer is able to make choices between the alternative uses and practices, which is appropriate in the USA mid-West where pressure on the land is light, but in the developing world the practical choices are much more limited. It is not reasonable to decree that land steeper than 12% is unsuitable for cultivation if that would take out half the cultivated land of the country. There is little opportunity for choice or selective use in countries with a food deficit, nor where the lack of alternative employment makes it necessary for everyone to maintain themselves from the land.

There are other constraints to applying this type of Capability Classification in developing countries, for example land is often not held on freehold tenure, and selective land use may not be practical under conditional or temporary tenure. Also in developing countries socio-economic conditions may override the physical conditions which determine the capability class. A recent study in Ethiopia illustrates the complexity of combining land capability and the peoples' needs. The approach suggested for regional planning is to first determine the technical options from Capability Classification and then select from these on the basis of a socio-economic analysis which is based on factors such as the shortage of food, shortage of fodder, and shortage of wood (HUISMAN and ESCOBEDO 1992).

Another limitation of the American system is that it does not cover all the possibilities which can arise in other situations. For example, in the humid tropics there can be deep productive soils on very steep slopes. Capability Classification would rate them unsuitable for cultivation on the grounds of slope, but after bench terracing and with good management they could be used for continuous cropping.

Procedures Since Capability Classification attempts to relate the use of land to the attendant risk of erosion, all the factors and characteristics which influence the risk of erosion must be considered and assessed. This is done by first collecting all the relevant facts in a survey, and then assembling them in a convenient order known as a *standard soil code*. The code has two purposes; it serves as a filing system, so that a particular piece of information is always stored in the same place where it can be quickly found. It also provides a convenient way in which a great deal of information can be recorded in limited space directly onto a map or aerial photograph.

The code consists of a series of letters and figures, each of which denotes the value of a particular characteristic. For example, if the effective depth of soil is more than 1 500 mm this is represented by the figure 1, between 900 and 1500 mm by 2, between 500 and 900 by 3, etc, and the figure representing soil depth is always placed in the same position at the top left-hand corner of the code.

A typical code would look something like this

$$\frac{2\,F\,5}{A\,1}$$

and the physical features described are in these positions

$$\frac{\text{Depth} - \text{Texture} - \text{Permeability}}{\text{Slope} - \text{Erosion}}$$

These are the main factors and are included in all coding systems, although each system has its own scale of values. The code is also a convenient way of separating those characteristics which relate specifically to the *soil*, and these are in the top line, whereas symbols relating to *land* characteristics are in the bottom line.

Some classification systems also allow for the addition of subsidiary factors which do not determine the capability class but may need to be recorded because they are useful information. These are shown to the right of an oblique stroke, thus

$$\frac{2\ F\ 5}{A - 1} \Big| \frac{2G}{Gs}$$

In this example from the Zimbabwe system the subsidiary information on the right of the oblique stroke is, on the top line, the colour of the upper subsoil (2) and the texture of the upper subsoil (*G*). Below the line the parent material (Gs) is the code for greenstone schists.

The criteria for capability classes The allocation of a piece of land into one of the eight capability classes is determined by considering several of the soil characteristics which have been assessed in the survey and recorded in the standard soil code. Each capability class has specified limits for each factor, and to be rated as belonging to a particular capability class the specification for every factor must be met. The essence of this classification is the risk of erosion, and this danger will be increased if the soil is vulnerable in any one respect, so classification depends upon the weakest factor, in the same way that the strength of a chain is that of its weakest link. The criteria for the four classes of arable land are usually much more detailed than for the non-arable classes, where a consideration of slope alone may determine the class. The criteria for two of the arable classes of the Zimbabwe system are shown in table 9.1.

Since some of the factors are more important than others in determining class, the process can be streamlined by considering the soil factors in a particular sequence as shown in figure 9.1. First the section of the diagram is selected according to the degree of slope, then the lines are followed down to the appropriate texture, then effective depth, then permeability. This leads to the first assessment of class, but the additional requirements (in the left column) must also be checked, and failure to comply with any requirement means downgrading to a lower class.

A variation of this method of allocating class by progressive rejection

Table 9.1 Criteria for arable land classes of the Zimbabwe classification

Land Capability Class	I	II	
Permissible slope	0–2%	0–2%	2–5%
Minimum effective depth (Texture here refers to average textures)	36 in. of CL or heavier	20 in. of Sal. or heavier	20 in. of SaCl or heavier
Texture of surface soil	CL or heavier	Sal or heavier S, or LS if upper subsoil is Sal or heavier	Sal or heavier
Permeability 5 or 4 to at least – Not worse than 3 to –	 36 in.	 20 in. 36 in.	 20 in. 36 in.
Physical characteristics of the surface soil. Permissible symbols.	Not permitted	t1	t1
Erosion— Permissible symbols	1	1 and 2	1 and 2
Wetness criteria— Permissible symbols	Not permitted	W1	W1

S = Sand,
Sal = Sandy Loam,
LS = Loamy Sand,
CL = Clay Loam
SaCL = Sandy Clay Loam,
From *Conservation Officers' Handbook*, Ministry of Agriculture, Zambabwe

was developed for quick assessment in the field and called *incapability classification* by SHAXSON (1981).

A classification for the humid tropics Where there is plenty of gently sloping land available for agriculture it is practical to impose restrictions on the maximum slope which is considered suitable for cultivation. That is the situation in the United States and in many of the other countries where LCC has been successfully applied. But in parts of the humid tropics the circumstances are different. Every hectare of land has to produce food for a hungry person and there is no reserve of undeveloped land. In these circumstances arable farming has to be extended to much steeper slopes by the use of various forms of terracing. A classification design for this

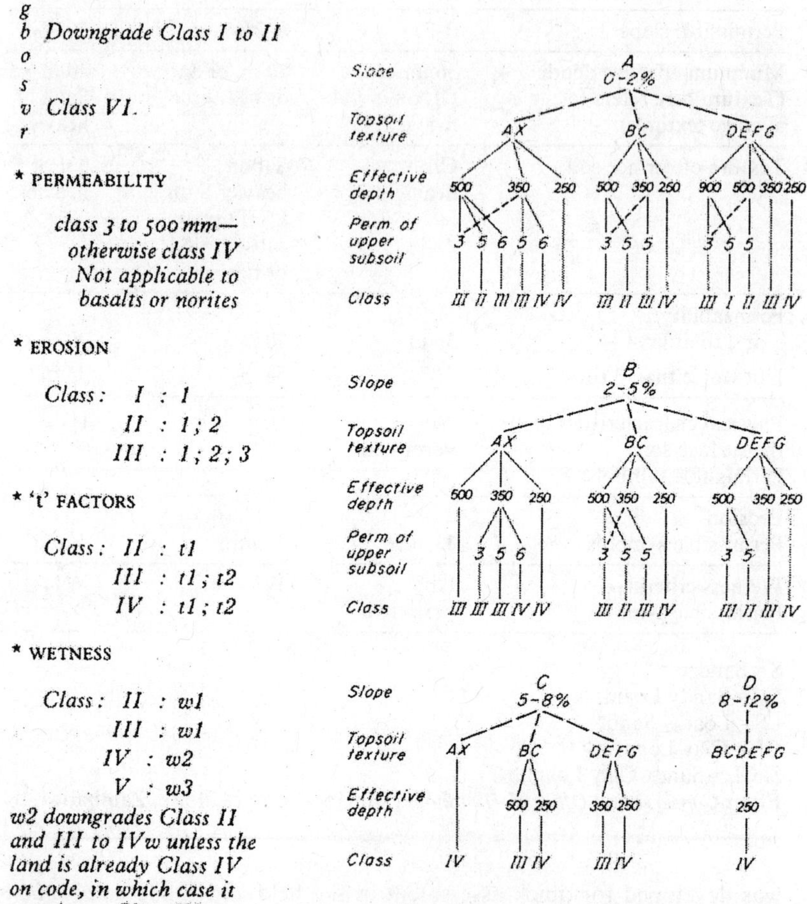

Figure 9.1 A chart for the systematic determination of capability class. Redrawn from a chart of the Planning Branch, Department of Conservation, Government of Zimbabwe

Table 9.2 Criteria for land capability classes in the humid tropics

Land Class	*Maximum slope degrees (Note 1)*	*Minimum soil depth (mm) (Note 2)*	*Conservation treatment (Notes 3, 4, 5)*	*Maximum intensity of land-use*
1	7		0–2 Contour cultivation	Any
			2–7 Channel terraces	Any
2	15	1000	Bench terraces	Any
3	20	500	Step terraces or hillside ditches	Close-cover crops and semi-perennials
4	25	500	Step terraces or hillside ditches	Tree crops with ground cover
5	33	250	Orchard terraces or platforms	Tree crops with ground cover (no cultivation)
6	More than 33	—	None	Forest only

Notes: 1 Equivalent slopes are 12, 27, 36, 42, 65 per cent.
2 Minimum soil depths are required when terraces are to be cut into the hillside.
3 The conservation treatment for Class 1 is the same as Classes I to IV in the SCS Land Classification System.
4 Terraces for Class 3 may be built and cultivated by small tractors.
5 Terraces for Class 4 will be built and cultivated by hand.

situation is shown in table 9.2 (SHENG 1989 and FAO 1976). The principle is the same as the original Land Capability Classification, that is to consider the physical soil and land conditions and then to recommend the most intensive form of production which is appropriate when combined with suitable conservation methods necessary to avoid undue erosion. In this case the two main considerations which determine the most suitable type of terracing are the slope of the land and the depth of the soil which are inter-related as discussed in section 10.5. Another difference is that this classification system allows for the production of non-food crops such as oil palm or rubber, and for tree crops such as tea, coffee, and spices.

9.1.6 Watershed planning

Some years ago planning and development on the basis of the watershed was very fashionable. Some countries channelled the whole of their conservation/extension efforts through this approach (as in Lesotho), others used it in addition to a national programme (India,

Kenya). There are significant advantages. Resources including money, equipment, and specialists can be concentrated in a limited area to achieve a greater impact than if these resources were more widely spread. This concentration probably makes it easier to attract the attention of the local population, although not necessarily their support. The disadvantages are that concentrating in one watershed means that other areas are deprived of resources, and also there is a danger that when the project moves on to the next watershed the support and interest dwindle.

Today opinion is divided. It is still strongly supported in India (BHARDWAJ and SINGH 1992) where it is a central feature of integrated watershed management through operational research projects. In Kenya a recent major survey claims that productivity has been significantly increased and sustained as a result of watershed planning (YERASWARQ ADMASSIE 1992). In Ethiopia the conservation/extension programme based on watershed planning collapsed almost overnight with the overthrow of the government in 1990 (HERWEG 1992). This approach of packing the watershed with specialists for a crash programme does not fit comfortably with the now generally accepted philosophy that all development should centre on encouragement and support for local self-help institutions and communities. There may be other disadvantages. In a recent study in Tanzania it is argued that: 'A technically sound approach like working on a catchment basis, may, in a village setting, in fact prove unrealistic or indeed harmful to the farmers. Soil conservation is a social process as much as a physical undertaking'. (CHRISTIANSSON *et al* 1992). Another problem is that the naturally occurring hydrological boundaries of watersheds seldom coincide with administrative boundaries. Provinces and Districts are more likely to be based on roads and access, and in rural areas a village is often sited on a crest with its lands sloping down on either side to the stream or river which is the village boundary.

9.2 New approaches to land use

9.2.1 From soil conservation to land husbandry

Land husbandry is the new brand name and so popular today that there are journal articles to define it (SHAXSON 1993), an excellent picture book for general consumption (LUNDGREN and TAYLOR 1993) and a serious text book (HUDSON 1992). There are several long and detailed definitions but I do not think it is necessary to expand beyond 'the care, management, and improvement of the land resource', with a possible side reference to the long established concepts of animal husbandry and crop husbandry. If the present trend continues 'soil conservation' may become a pejorative term as has happened to 'conservation engineering'.

Land husbandry is not really new, but there seldom is anything new in land use – see Virgil's *The Georgics* Book I on conservation farming or the account of Roman conservation works in north Africa (LOWDERMILK 1975). The components of today's land husbandry have been around for a very long time. The soil chemists know about the carbon/nitrogen ratio, plant nutrient requirements and so on. The soil physicist knows about compaction and aeration, porosity and permeability, and so on. The conservation engineer knows about increasing infiltration and moisture storage capacity to reduce runoff, and managing the runoff when it cannot be avoided, and so on. Every farmer and gardener knows what it means for soil to be 'in good heart'.

The surprising thing is that it has taken us so long to put all the parts of the jigsaw together and package it under a good brand name. There was a severe reaction when in 1987 I wrote that there ought not to be any soil conservation departments because soil conservation should be an integral part of extension, a view influenced by the highly successful combined department of conservation and extension (CONEX) which in Rhodesia (now Zimbabwe) 50 years ago set the example for Africa (HUDSON 1988). The introduction of the land husbandry approach to land use has also brought about a realignment of our thinking about the relationship between mechanical erosion control measures and what, for want of a better term, we lump together as biological control measures. The old approach derived from our experience of applying soil conservation practices to 'western-style' commercial farming where the idea was first to install appropriate mechanical works such as terraces and drains, and then follow up with improved farming practices on the protected land. This was, and still is, a sound approach for large-scale mechanized farming, but we made the mistake of trying to extend this plan to the very different conditions of small-scale farming. In chapters 10 and 11 we will discuss control measures. The point to make here is that the new land husbandry approaches reverse the priority, and says 'let us first do all we can to reduce erosion by improving crop management and productivity, and use mechanical controls to reinforce biological control only when necessary'. These ideas are developed more fully in SHAXSON *et al* (1989).

9.2.2 *People participation*

The involvement and participation of the population is nowadays proclaimed to be the cornerstone for building agricultural development, but this also is not so much a fundamentally new approach as a further development of established practices. The father of soil conservation said:

> 'National action to conserve the soil resource must not only deal with soil erosion and related physical phenomena but also cope with the complex economic and social considerations that affect land use.
>
> 'The ultimate objective of national action – a secure land resource, adequate permanently to sustain the national economy – will be reached only when the principles of good land use have been accepted and put into practice by agriculture as a whole.
>
> 'But agriculture as a whole is based on the activities of individuals. The land is owned and used by many men. They enjoy the rights and privileges of private ownership under the American system. In this democracy national action to conserve soil must be generated by these millions of land users. If they are active and willing participants in such a movement, it will endure; otherwise it will fail. Democracy does not give government the unbridled right to compel conservation; but it does recognise the responsibility of government to lead and assist its citizens in conservation, in the interest of the public welfare.' (HUGH HAMMOND BENNETT 1939)

A basic principle of many extension services is that the field officer should live and work among and with the farmers and so is completely familiar with their practices, social structure, problems and aspirations. The need for surveys and appraisals to find out what the locals are doing and thinking only arises when the extension workers are outsiders following the old-fashioned idea that all that is necessary is to show the poor ignorant peasants what they should be doing, and if necessary, tell them what to do for their own good.

These excellent precepts have not always been followed and we might look at two examples. First it is true that in the days of colonial soil conservation we went off the rails for a time when we thought that all we had to do was to show farmers what they ought to be doing. We would not use this approach nowadays, but it does not deserve the excessive condemnation which it sometimes receives. A list of British Colonial territories in east and central Africa which had strong active conservation programmes comprises Southern Rhodesia, Northern Rhodesia, Nyasaland, Kenya, Tanganyika, Bechuanaland, Basutoland, and Swaziland. If we now list the countries that today have the most advanced natural resource policies we find, Zimbabwe, Zambia, Malawi, Kenya, Tanzania, Botswana, Lesotho, and Swaziland. Can it be coincidence that the only difference is that some of the names have changed?

The second example of attempting conservation and extension without true involvement of the farmers was the extension system taken up and strongly supported by the World Bank a few years ago but now largely abandoned. It was called the *Training and Visits Programme* and was based on the presumption that there would never be sufficient extension workers in developing countries to be able to make contact with all the farmers, and that the last stage of dissemination should be from the extension worker to selected lead farmers who would in turn pass on the message to their own group of 12 to 20 local farmers. It was a tightly scheduled process with some days set aside for the training of the extension workers and other days for them to go round to their groups of lead farmers on the same day at the same time every alternate week. The basic weakness was that this was purely one-way communication, with no feedback about what the farmers were trying to do or wanted to know.

Lack of involvement and participation by farmers was found to be one of the most frequent causes of failure of most soil conservation projects (FAO 1991), but the lesson appears to have been well learned now. The rapidly growing literature divides into three groups representing successive phases. First there were studies and workshops in many countries on how to achieve popular involvement, for example early reports from FAO (1983a, 1983b). A workshop in 1987 at the Institute of Development Studies, University of Sussex, led to the classic book on the subject *Farmer First* (CHAMBERS *et al* 1989). Nepal was the centre of studies of people participation in the upland watersheds of the Hindu Kush-Himalaya region covering Afghanistan, Pakistan, India, China, Bangladesh, and Burma (DANI and CAMPBELL 1986). The proceedings of several workshops in southern Africa have been published (SADCC 1987 and 1991).

The second group is studies of farmers' attitudes to innovation and adoption and the constraints which can hold it back. The message which comes out clearly from these studies is that the constraints are very complex and often difficult to identify. The farmers themselves are not always clear. For example in a study of hill farmers' approach to soil conservation in northern Thailand the farmer says that the main constraints to adopting soil conservation were lack of labour and lack of money, but they also held the belief that erosion was not serious so it is not clear which was the dominant factor (HARPER and EL-SWAIFY 1988). In a detailed analysis of implementation in Indonesia, Lok groups the constraints under the headings lack of skills or knowledge, low income, ineffective communication, and insecure tenure (LOK 1992).

A common theme in developing countries is the lack of experience of self-management through local institutions. Particularly in Africa many countries have a history of autocratic rule whether by colonial powers, kings, tribal chiefs, or socialist parties, and institution-building is nowadays increasingly a component of development plans. In developed countries the barriers to people participation in conservation plans are equally complex and not very different from developing countries. For example in Spain the thought is reported that erosion is not very serious (EPPINK 1992) and in New South Wales, Australia, the reasons suggested are that the proposed measures are too complex, require too much capital and intellectual outlay and that the risk associated with the new techniques is unacceptable (VANCLAY 1992).

The third group of literature is reports of successful projects involving people in planning their own future, and here are rich seams to be mined. Keying into a reference retrieval system either 'sustainability?' or 'people participation?' will result in a flood of conferences, books, and journal papers all from the last five years. The best starting point is *The Greening of Africa* (HARRISON 1987), probably the first breakout from the scenario of doom and gloom in Africa. This was closely followed by *The Greening of Aid* (CONROY and LITVINOFF 1988) and *Towards Sustainable Development* (PANOS 1987), both deriving from conferences which discussed successful projects as well as the problems. An excellent series of case studies has been sponsored and reported by the World Resources Institute (see *Further reading*), and many others were discussed at a workshop in Kenya and Tanzania (HUDSON *et al* 1993).

Africa has been the main focus of this movement, probably because the problems of soil degradation and food supply are particularly severe (BROWN and WOLF 1985), but the international exchange of new ideas is now so well developed through networks and international conferences that the same trends are appearing in every continent as can be seen from the proceedings of recent International Soil Conservation Organization conferences (see *Further reading*). This is clearly going to be seen as the way forward in the rest of this decade, and the trend to participatory programmes and projects will accelerate dramatically as the big international aid agencies follow the lead of the NGOs.

9.3 Traditional land use practices

Another currently topical enthusiasm is the study of traditional farming methods, and there is good evidence that new practices or methods are more likely to be accepted and adopted if they are improvements on practices which are already familiar. There are now twelve regional centres for the study of indigenous technology, including agricultural technology, with an international coordinating office, and national and international conferences (CIRAN 1993).

As with all conservation methods we must be careful not to extrapolate traditional methods nor to expect them to work outside the conditions in which they were developed. We must also be careful not to assume the reasons why a practice has been used, for the farmers aim is seldom a single objective like saving water or improving yield. A number of the practices presently listed among 'traditional soil conservation methods' are in fact primarily good land husbandry practices which happen to be helpful in reducing erosion.

Bench terracing on steep slopes is the practice most likely to have been done specifically for soil conservation, but even in this case we cannot tell how important the control of erosion was compared with other benefits like easier cultivation, better yields from improved soil moisture, and counteracting a rundown of fertility. Similarly, in the case of manuring, it is probable that the obvious fact that it improves yield is the main reason for adoption, and the fact that a well-manured crop gives better cover and reduces splash erosion trails a long way behind in the farmer's reasoning.

In a recent paper MILLINGTON (1993) points out that the clear advantage of recognizing and building on traditional practices should be tempered by the fact that they may not necessarily be appropriate in current conditions. An example of this is reported from Kenya, where pressure of population in the highlands is forcing migration down into the drier lowlands. The farming practices of the high-potential highlands are not appropriate for the poorer soils and lower rainfall, but some farmers are having difficulty in abandoning their traditional practices, and readjusting to the new conditions (LINIGER 1992).

There can be other reasons for old-established methods to become inappropriate. One is the lack of the large labour force required to build new terraces or maintain the old ones, and a well-documented example of this comes from the Yemen Arab Republic. Terraced land which has been cropped for 300 years with a high density of population, is now being abandoned as a result of massive migration to the oil-rich economies of Saudi Arabia and other Arab Gulf states (VOGEL 1987). Other social or economic exchanges may upset traditional methods, for instance, land allocation may introduce new styles of land tenure; the increase of livestock numbers may require different management techniques of communal grazing land; or the change from subsistence farming to a market economy may reduce the need for traditional techniques based on the need for survival. Useful compilations of traditional agricultural practices in Africa are MCCALL (1988), REIJ (1991) and CRITCHLEY *et al* (1994). A world wide review is PAWLUK *et al* (1992). Some of the traditional farming practices which have a significant soil conservation component will be discussed in chapters 10 and 11.

References

ABRAHAM, S. 1992 Land protection into the future using the soil data system, in *People Protecting their Land*, edited by P. G. Hoskins and B. M. Murphy, Department of Conservation and Land Management, Sydney, New South Wales, Vol 1, 705–710

BENNETT, H. H. 1939 *Soil Conservation*, McGraw-Hill, chapter XV, 313

BHARDWAJ, S. P. and G. SINGH 1992 Achievement of operational research projects on integrated watershed management for sustained productivity, in *People Protecting their Land*, edited by P. G. Hoskins and B. M. Murphy, Department of Conservation and Land Management, Sydney, New South Wales, Vol 2, 524–532

BROWN, L. R. and E. C. WOLF 1985 Reversing Africa's Decline, *Worldwatch Paper 65*, Worldwatch Institute, Washington DC

CHAMBERS, R., A. PACEY and L. A. THRUPP (editors) 1989 *Farmer First: Farmer Innovation and Agricultural Research*, Intermediate Technology Publications, London

CHRISTIANSSON, C., W. OSTBERG, V. M. LOISKE and C. LINDBERG 1992 History, society, and local production systems, Crucial variables in modern soil conservation work, in *People Protecting their Land*, edited by P. G. Hoskins and B. M. Murphy, Department of Conservation and Land Management, Sydney, New South Wales, Vol 2, 384–390

CIRAN 1993 *Indigenous Knowledge and Development Monitor*, CIRAN, Box 90734, 2509 LS The Hague, Netherlands

CONROY, C. and M. LITVINOFF (editors) 1988 *The Greening of Aid*, Earthscan, London

CRITCHLEY, W. R. S., C. REIJ and T. J. WILLCOCKS 1994 Indigenous Soil and Water Conservation: a review of the state of knowledge and projects for building on traditions, *Land Degradation and Rehabilitation*, 27, special issue

DANI, A. A. and J. G. CAMPBELL 1986 Sustaining upland resources. People's participation in watershed management, *Occasional Paper 3*, ICIMOD, Nepal

EPPINK, L. A. A. J. 1992 Groups involved in planning and implementing soil conservation measures – A case study from the Conca de Tremp area, Spain, in *People Protecting their Land*, edited by P. G. Hoskins and B. M. Murphy, Department of Conservation and Land Management, Sydney, New South Wales, Vol 2, 541–548

FAO 1976 A framework for land evaluation, *Soils Bulletin 32*, FAO, Rome

FAO 1978 Report on the Agro-Ecological Zones Project, Vol 1, *Methodology and Results for Africa*, Vol 2, *Results for Southwest Asia*, World Soil Resources Report 48, FAO, Rome

FAO 1983a *People's Participation Projects: guiding principles on design, operation, maintaining, and ongoing evaluations* by G. Huizer, Human Resources, Institutions and Agrarian Reform Division, FAO, Rome

FAO 1983b *Training of Group Promotors in field projects of the people's participation programme*, Human Resources, Institutions and Agrarian Reform Division, FAO, Rome

FAO 1991 A study of the success or failure of soil conservation projects, *Soils Bulletin 64*, FAO, Rome

HARPER, D. E. and S. A. EL-SWAIFY 1988 Sustainable agricultural development in North Thailand: conservation as a component of success in assistance projects, in *Conservation Farming on Steep Lands*, edited by W. C. Moldenhauer and N. W. Hudson, Soil and Water Conservation Society, Ankeny, Iowa

HARRISON, P. 1987 *The Greening of Africa*, Paladin, London

HERWEG, K. 1992 Major constraints to effective soil conservation – experiences in Ethiopia, in *People Protecting their Land*, edited by P. G. Hoskins and B. M. Murphy, Department of Conservation and Land Management, Sydney, New South Wales, Vol 2, 404–412

HUDSON, N. W. 1988 Tilting at windmills or fighting real battles, in *Conservation Farming on Steep Land*, edited by W. C. Moldenhauer and N. W. Hudson, Soil and Water Conservation Society, Ankeny, Iowa

HUDSON, N. W. 1992 *Land Husbandry*, Batsford, London

HUDSON, N. W., R. J. CHEATLE, A. P. WOOD and F. GICHUKI (editors) 1993 *Working with Farmers for Better Land Husbandry*, Intermediate Technology Publications, London

HUISMAN, P. and J. ESCOBEDO 1992 Land Capability versus Socio-economic Conditions: A Challenge for Soil Conservation in the Ethiopian Highlands, in *Soil Conservation for Survival*, edited by H. Hurni and K. Tato, Soil and Water Conservation Society, Ankeny, Iowa, Vol 1, 521–529

HURNI, H. 1986 *Soil Conservation in Ethiopia*, Department of Community Forests and Soil Conservation, Ministry of Agriculture, Ethiopia

KLINGEBIEL, A. and P. H. MONTGOMERY 1961 Land Capability Classification, *Agriculture Handbook 210*, USDA-SCS, Washington DC

LINIGER, H. P. 1992 A study of the effects of land use on water and soil resources on the slopes of Mount Kenya, in *Working with Farmers for Better Land Husbandry*, edited by N. W. Hudson and R. J. Cheatle, Intermediate Technology Publications, London, 208–211

LIU, Z., G. O. EYLES and P. F. NEWSOME 1992 GIS as the basis for preparing soil conservation farm plans, in *People Protecting their Land*, edited by P. G. Hoskins and B. M. Murphy, Department of Conservation and Land Management, Sydney, New South Wales, Vol 2, 723–731

LOK, S. H. 1992 People's Participation in Conservation Farming, in *Conservation Policies for Sustainable Hillslope Farming*, edited by S. Arsyad, I. Amien, T. C. Sheng and W. C. Moldenhauer, Soil and Water Conservation Society, Ankeny, Iowa

LOWDERMILK, W. C. 1975 Conquest of the Land through seven thousand years, *Agriculture Information Bulletin 99*, USDA-SCS, Washington DC

LUNDGREN, L. and G. TAYLOR 1993 *From Soil Conservation to Land Husbandry*, Natural Resources Management Division, SIDA, Stockholm

MCCALL, M. K. 1988 Indigenous Technical Knowledge in farming systems and rural technology: a bibliography on eastern Africa, *Working paper 38*, Technology and Development Group, University of Twente, Enschede, Netherlands

MILLINGTON, A. C. (editor) 1993 Indigenous soil and water conservation and harvesting, *Proceedings of E.C. Conference at Chania, Crete, 1991*, Penny Press, Donnington, England

PANOS 1987 Towards Sustainable Development, *Nordic Conference on Environment and Development*, Panos Institute, London

PAWLUK, R. R., J. A. SANDOR and A. TABOR 1992 The role of indigenous soil knowledge in agricultural development, *Journal of Soil and Water Conservation*, 47, 4, 298–302

REIJ, C. 1991 Indigenous soil and water conservation in Africa, *Gatekeeper Series 27*, IIED, London

SADCC 1987 People's participation in soil and water conservation, *Report 10, SADCC Soil and Water Conservation and Land Utilization Programme*, Maseru, Lesotho

SADCC 1991 Promotion of people's participation in land utilization, *Report 26, SADCC Environment and Land Management Sector*, Maseru, Lesotho

SHAXSON, T. F. 1981 Determining erosion hazard and land use capability – A rapid subtractive survey method, *Soil Survey and Land Evaluation*, 1, 3, 44–50

SHAXSON, T. F. 1993 'Land Husbandry' and 'Soil Conservation', *Splash*, 9, 2, 5

SHAXSON, T. F., N. W. HUDSON, D. W. SANDERS, E. ROOSE and W. C. MOLDENHAUER 1989 *Land Husbandry: A Framework for Soil and Water Conservation*, Soil and Water Conservation Society, Ankeny, Iowa

SHENG, T. C. 1989 Soil Conservation for small farmers in the humid tropics, *Soils Bulletin 60*, FAO, Rome

STREET, G. J. and A. C. DUNCAN 1992 The application of airborne geophysical surveys for land management, *People Protecting their Land*, edited by P. G. Hoskins and B. M. Murphy, Department of Conservation and Land Management, Sydney, New South Wales, Vol 2, 762–770

THOMPSON, B., K. MCPHEE, J. SALMON, N. SARGENT and R. TUCK 1992 Changing Land Management with Property Planning, in *People Protecting their Land*, edited by P. G. Hoskins and B. M. Murphy, Department of Conservation and Land Management, Sydney, New South Wales, Vol 2, 771–775

TOSI, J. A. 1964 Climatic Control of Terrestrial Systems – A report on the Holdridge Model, *Economic Geography*, 40, 2, 173–181

VANCLAY, F. 1992 The barriers to adoption often have a rational basis, in *People Protecting their Land*, edited by P. G. Hoskins and B. M. Murphy, Department of Conservation and Land Management, Sydney, New South Wales, Vol 2, 452–453

VELOZO, R. DE C. 1985 Incentives for participation of the community in soil conservation programmes, *Conservation Guide 12*, FAO, Rome

VIRGIL, *The Georgics*, Book 1, translation into English by L. P. Wilkinson 1982, Penguin Books, London

VOGEL, H. 1987 Terrace Farming in Yemen, *Journal of Soil and Water Conservation*, 42, 1, 18–21

YERASWARQ ADMASSIE 1992 *The Catchment Approach to Soil Conservation in Kenya*, Regional Soil Conservation Unit, SIDA, Nairobi, Kenya

ZHOU, Q. 1989 A Method for Integrating Remote Sensing and Geographic Information Systems, *Photogrammetric Engineering and Remote Sensing*, 55, 5, 591–596

10

Physical erosion control measures

10.1 Different practices for different purposes

There are so many different practices used in erosion control that to describe them needs some form of grouping. For a long time *mechanical protection* was used to describe all the methods which involved earth moving, such as digging drains, building banks, levelling sloping land, and so on. Anything else was lumped together under *biological methods* for want of a better title.

This is appropriate for large mechanized farms where the approach is to use machines to do the earth-moving and follow up with improved farming methods. But this division does not suit the concept of erosion control through better land husbandry supported by mechanical protection, and the division is artificial when we start talking about progressive terracing using grass strips or live hedges.

An approach more suited to today's thinking is to group methods and techniques according to their main objective. All have an element of saving soil loss, but an important difference is the handling of surface runoff. This chapter identifies three kinds of water management objective and two types of soil management on sloping land.

- In 10.2 the objective is to hold all the rain where it falls, or to transfer additional runoff onto arable land.
- In 10.3 the objective is to retain as much rainfall as possible, but to cater for infrequent overflows.
- In 10.4 the objective is to manage unavoidable runoff.

There are no precise divisions between structures built with the main objective of holding all the rainfall (perhaps with an emergency overflow for exceptional storms), and therefore discussed in 10.2, and structures built to hold part of the rainfall and release some through a controlled drainage system and therefore included in 10.3.

Slope-management practices are divided into:

- 10.5 slope modifications by 'one-shot' terracing, and
- 10.6 progressive slope modification over time.

Erosion control through crop management is discussed in chapter 11, and control measures for non-arable land in chapter 13. There is bound to be some overlap across these arbitrary divisions because practices have different objectives in difficult circumstances. Terracing may be for erosion control or water retention, or both; water harvesting may be to improve grazing, or to allow limited cropping, or to assist tree planting. We have used cross-references liberally to avoid describing a practice in more than one place.

It must be emphasized that none of the methods described have universal application. In selecting from possible methods several issues must be considered.

- Any method must be suitable for the intended land use and cropping systems.
- The objectives must relate to rainfall and soil. In high rainfall areas a common objective is to lead unavoidable surface runoff safely off the land using drains and ditches. In semi-arid regions the objective is more likely to be to slow down the runoff to non-scouring velocities to encourage infiltration or deposition of silt.
- The inputs, especially of labour, must be affordable and the benefits must be sufficient to justify the inputs.

Table 10.1 summarizes the practices discussed in this chapter, and figure 10.1 relates types of terrace to land slope. The naming of various kinds of earth works and structures used to control erosion is complicated because in some cases the same structure is called by different names in different countries, in other cases the same name is applied to different things. Alternative names are given wherever possible.

10.2 Water management – maximizing retention

10.2.1 Suitable conditions

Erosion will be almost eliminated and crop production will in most cases be improved if all the rainfall is absorbed and retained in the soil. There must be sufficient moisture storage capacity in the soil, and this is a function of soil depth, structure, porosity, and organic matter.

Main Objective	*Function*	*Type of terrace or barrier*
Soil Management	To modify soil slope	Bench terraces, plates 10.15, 10.16 and 9.1
	To slowly reduce soil slope	Progressive terracing, figures 10.13 and 10.14
	To contain erosion with low inputs	Ladder terraces, plate 10.28 Trash lines, plate 10.7
	To contain erosion with minimal earth-moving on steep slopes	Step terraces, plate 10.19 Hillside ditches, plate 10.23 and 10.24 Intermittent terraces, plate 10.20
Water Management	To multiply effective rainfall	Conservation bench terraces, figure 10.2 Runon level terraces
	To catch and hold all the runoff	Absorption ridges, plate 10.1
	To absorb some run-off with emergency overflow	Contour furrows, plate 10.4 Contour bund, plate 10.5
	To control unavoidable runoff	Graded channel terraces, plate 10.10
	To control reduced run-off	Ridging, plates 10.7, 10.13 and 10.14 Tied-ridging, plate 10.9
	To reduce the velocity of runoff and promote infiltration	Strip cropping Grass strips, plates 10.29, 10.30, 10.32 and 10.33 Permeable barriers
Crop Management	To provide level areas on steep slopes, or to ease cultivation according to whether by hand, ox, or machine	Step terraces, plate 10.19 Hillside ditches, plates 10.23 and 10.24 Orchard terraces, plate 10.20 Platforms, plates 10.21 and 10.22
	To ease harvesting according to whether the crop is heavy, damageable, harvested regularly or seasonally	Footpaths and farm tracks associated with orchard terraces or hillside ditches, plate 10.20
	Drainage for crops which suffer from wet feet	Ridges on 2% grade for tobacco Up and downslope beds for yams Small open drains up to 15% for teff, plate 10.12

Table 10.1. *Terraces and other cross-slope barriers for different purposes*

Classifying conservation works according to their purpose cannot be exact because most have more than one objective – for example any cropping practice which improves water availability could be listed under crop management or water management. Any device which traps and holds soil will also catch water (figure 10.1, opposite)

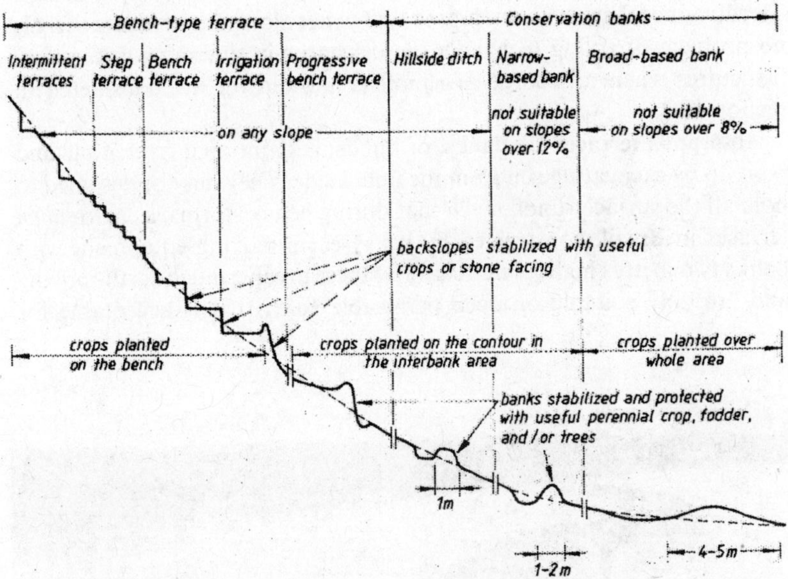

Figure 10.1 Types of terrace for different slopes

Methods to improve infiltration, storage on the surface, and in the soil, are discussed in chapter 11.

There will be situations where maximizing infiltration is not practical or not desirable, for example shallow soils, soils with poor drainage, or crops which suffer from restricted drainage in the root zones, or where mass movement by landslips or slides is a problem. But in semi-arid regions, or regions with moderate rainfall and good soils, maximizing rainfall retention is usually desirable.

10.2.2 Retention methods for arable land

Structures on the contour are simpler and cheaper than structures on grade to discharge runoff, for three reasons. Firstly, there is no need to set them out on a precise gradient. They should be more or less on a level contour, but small errors are not as important as in the case of graded channels. Secondly, where water is to be led off the land, then the spacing between the terraces has to be calculated because each channel terrace has to handle the water from a given area. If the object of structures on the contour is to store the total runoff then they must be designed to do this, but the distance between them is dictated by their size. Thirdly, since there is no attempt to lead water along the structure there is no problem of trying to handle the discharge in drains or waterways. Structures where discharge of runoff is allowed for are considered in section 10.3.

Absorption terraces are banks, or bunds, laid out on a level grade and built up by excavating soil from the uphill side. They may be designed to hold all the surface runoff, even that during heavy storms, as *murundum* terraces in Brazil, constructed by heavy earth-moving equipment with banks two metres high (plate 10.1), but these involve much earth-moving and are only suitable on deep permeable soils. If it is acceptable for

Plate 10.1 Murundum terrace in Brazil, intended to hold all surface runoff, even during heavy storms (T. F. Shaxson)

Plate 10.2 Contour bunds are intended to catch both soil and water

exceptional storms to discharge through emergency spillways, the banks can be lower, as in the contour bunds of central India with banks 1.5 metres high and spillway crests at 1 metre (plate 10.2).

Irrigation terraces Flat bench terraces are most frequently built without a specific objective of either draining away surface runoff or trying to hold it on the terrace. The exception is irrigation terraces where there is a raised lip at the outer edge to retain irrigation water, as extensively used for the production of rice, and to a lesser extent for tea, fruit trees, and other high-value crops. For paddy, the terraces are also level along their length, so that each becomes a flooded shallow pond, but where intermittent water application is intended there can be a slight longitudinal fall. The Banaue rice terraces in the Philippines are a magnificent example of irrigation terraces, a monumental engineering feat of great antiquity, but still maintained today (plate 10.3). This type of terrace is found in most regions where rice is the staple food, and is very common in Indonesia, Malaysia, China, and Japan.

Rainfall multipliers This term is applied to methods where the runoff from an uncultivated part of the land is diverted onto a cultivated part, thus giving it the benefit of more water than it receives as rain, hence the name rainfall multiplier. When designing such methods the object is to get the best ratio of the area yielding runoff to the area receiving runon.

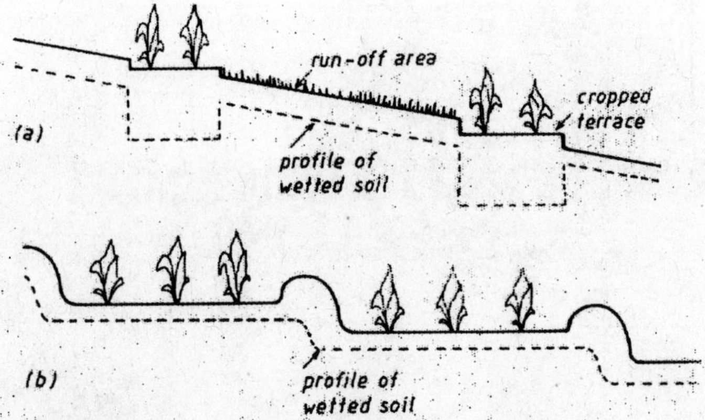

Figure 10.2 The effect of conservation bench terraces (CBTs)
(*a*) Increased available soil moisture under CBTs
(*b*) Level bench terraces do not have a rainfall multiplier effect like CBTs

Because of variations of soil and crop requirements, and above all the variability of rainfall, this cannot be a straightforward mathematical calculation and must be established by field trials. A system used in the drier southwest of the United States of America for large scale mechanized cropping is called *conservation bench terraces* and the effect on available moisture is illustrated in figure 10.2.

The main features required for conservation bench terraces are:

- Gentle slopes of 0.5–1.5% are most suitable.
- A deep soil is required, both to provide sufficient soil moisture storage, and also to lessen the effect of cutting down during the construction of the terraces. Good permeability is required so that the contained runoff can be absorbed quickly.
- Smooth slopes are an advantage where mechanized farming can be made more convenient by constructing all the terraces parallel and of equal width.
- Precise levelling of the bench terrace is important to ensure uniform infiltration.
- If there is a risk that runoff from the catchment area will be

Plate 10.3 The Banaue irrigation rice terraces of the Philippines, originally built many centuries ago and still intensively used. (Philippine Tourist and Travel Association)

Plate 10.4 Contour furrows in north-eastern Brazil (H. Lal)

greater than can be absorbed and stored on the terrace there must be outlets at the end of the terrace.

- Typical ratios of catchment area to terrace area are from 1:1 to 2:1.

Contour bunds, also known as *contour furrows* or *desert-strip farming.* These are variations of rainfall multiplier which require less soil movement than conservation bench terraces. The cropping is usually intermittent on strips or in rows, with the catchment left fallow (plate 10.4). Studies of the effect on soil moisture have been made in Kenya (SMITH and CRITICHLEY 1983 and figure 10.3) where a crop of sorghum and cow peas (*Vigna sinensis*) was grown on only 270 mm of rainfall.

A disadvantage of contour bunds is that the crop tends to be uneven, reflecting the soil moisture profile, and also only a small part of the field is cropped. But the point is that the method enables some crop to be grown where otherwise the rainfall would be inadequate for any cropping. A variation is where the whole soil surface is shaped, as shown in figure 10.4.

Another approach to localized surface storage on part of the land is the use of semi-circular catchments or trapezoidal structures (figure 10.5 and plate 10.5). The large number of separate semi-circular or trapezoidal bunds spreads the risk of damage from overflow in exceptionally heavy rain, or if one bund fails. A similar method is used to improve water retention in low rainfall areas, as discussed in sections 13.1.2 and 13.4.

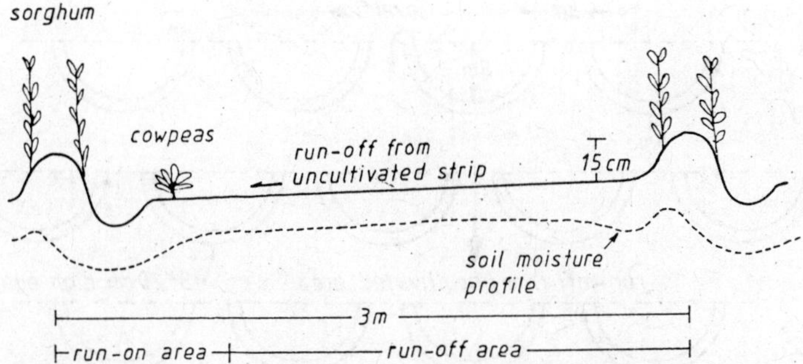

Figure 10.3 Improved soil moisture from contour bunds (SMITH and CRITCHLEY 1983)

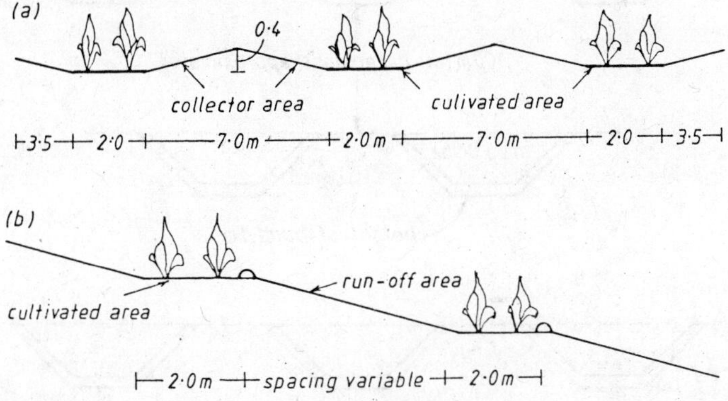

Figure 10.4 Shaping the whole surface into runoff and runon areas: (*a*) On level ground (*b*) On sloping ground

10.3 Water management – limited retention and some drainage

10.3.1 Suitable conditions

Retaining all the rainfall may not always be practical or desirable, so there is a group of practices which are able to retain part of the rainfall and to manage or control the remaining runoff. This is a common condition when storms are likely to occur occasionally, and it would not be practical economics to apply the principle of the Brazilian murundums (section 10.2.2) which are huge expensive structures expected to hold all runoff, or the Indian contour bunds, expected to hold the runoff in most storms. This group also includes practices which temporarily slow down runoff and so give it more time to infiltrate.

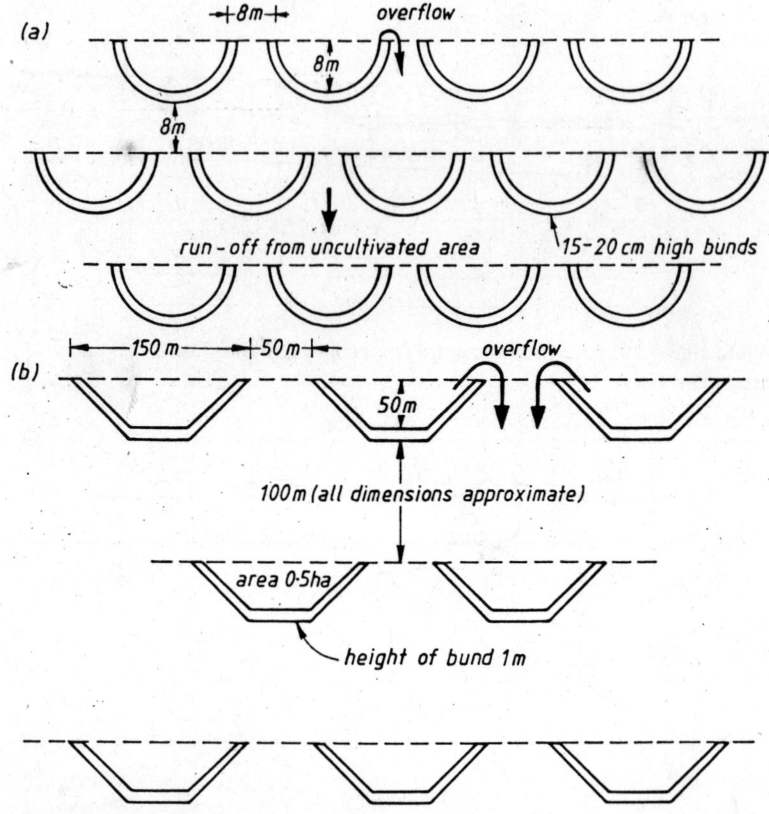

Figure 10.5 Simple structures to retain surface runoff:
(*a*) Semi-circular hoops (SMITH and CRITCHLEY 1983)
(*b*) Trapezoidal catchpits (FINKEL 1986)

10.3.2 Practices for arable land

Contour cultivation On gentle slopes or whenever the erosion risk does not warrant major earth-moving works, it may be sufficient to slow down surface runoff by carrying out all tillage operations on the contour. This eliminates furrows and depressions running down the slope which invite runoff to concentrate in them with the result of causing scour erosion.

Farming on a grade is well established in India (SWAMINATHAN 1982). Cultivations and planting are done on a gentle gradient which encourages infiltration but permits surplus runoff at low velocities. Sometimes this

Plate 10.5 Semi-circular hoops catch and retain surface runoff, Niger (C. Reij)

may be combined with simple practices to encourage infiltration such as returning crop residues. There can be still be a problem of disposal of the surface runoff when it does occur.

Strip cropping is planting along the contour alternate strips of a row crop, such as maize or sorghum, with alternate strips of a forage crop. This is a popular and effective method for large farms on gentle slopes which are farmed in a rotation. Instead of putting each course of the rotation on the whole field, the successive courses are planted successively on each strip, so that each year the rotation marches down the field by one strip width. This is sometimes described as rotation by space instead of by time.

Ridging In all forms of cultivating on the contour, tillage, weeding, and other operations cause the formation of small banks and ridges that impede the downslope flow of runoff, and improve the infiltration. The amount of surface storage can be enhanced by deliberately building up continuous ridges. An example of large ridges built by machine is shown in plate 10.6 which is on a tea plantation in Malawi; plate 10.7 shows simple ridges built up from soil and crop residues in Ethiopia. If the ridges are laid out on the true contour there is a danger that they may overtop in a heavy storm and this is likely to cause damage through the knock-on effect in a manner similar to that when channel terraces fail as discussed in

Plate 10.6 Ridging on a tea experiment station in Malawi (T. F. Shaxson)

section 10.4.2 and plate 10.11. One solution is to build the ridges on a gentle gradient and allow them to discharge into a natural or improved water course or into a designed drain.

Tied-ridging An improvement on simple ridging, where it suits the cropping system, is to build cross ties between the ridges as shown in plate 10.8. Two safety features are required to cope with the exceptional storm. The ridges should be on grade, and the ties should be lower than the ridges so that if there is any overtopping it will be along each furrow and not down the slope.

Tied-ridging is usually associated with mechanized farming, but there has been some success in achieving it with ox-drawn implements as illustrated in plate 10.9. There is considerable evidence that tied-ridging is an effective method of increasing crop yield, particularly in regions of low or unreliable rainfall, and particularly with crops which will not be badly effected by the temporary saturation of the furrows (HULUGALLE 1988). Most studies stress that there must be local field trials using different crops and variations such as whether the crop is planted in a bottom of the furrow or in the ridge. The question of the additional labour for building

Plate 10.7 Simple ridges and ties built up from soil and crop residues, Ethiopia (H. Hurni)

ridges and ties is sometimes a deterrent and a system to reduce the labour requirement is being developed in Zimbabwe using ox-drawn implements and called *no-till tied-ridging* (ELLWELL and NORTON 1988). In the first year the land is ploughed, then ridges are built on grade by an ox-drawn ridger, and cross-ties built by hand or an ox-drawn scoop. In subsequent years the land is not ploughed and the ridges and ties need only to be maintained at the proper size and shape. There is a high labour requirement in the first year, but little in the following years. If ploughing and ridging is repeated after four or five years, the total labour input over the cycle is low. Experiment station trials have shown that the system is effective, and after well-managed farmer trials the system is being widely adopted (NYAMUDEZA *et al* 1993).

Permeable barriers are usually intended to catch and retain both water and soil, and are discussed in section 10.6.2 for arable land, and section 13.1.3 for grazing land.

10.4 Water management – controlled runoff

10.4.1 Suitable conditions

Situations can arise where it is appropriate to use channels based on hydraulic design. Sometimes it is predictable that considerable volume

Plate 10.9 A simple ox-drawn tool to make intermittent ties, Zimbabwe (H. Elwell)

of runoff will occur frequently, for example if the soil is shallow and liable to become saturated so that the rainfall cannot infiltrate. Another case is when the intensity of rainfall exceeds the surface infiltration rate so water ponds on the surface and runoff occurs, sometimes called *Hortonian flow*. In some situations it may be appropriate to encourage some controlled runoff either because there is a risk of the soil becoming waterlogged, or because the crop 'can't stand wet feet', such as tobacco or yams.

10.4.2 The classic design of mechanical protection works

This system was developed by the USDA-SCS 50 years ago and is still relevant in the circumstances for which it was designed, that is large arable lands on large mechanized farms where the design is provided by soil conservation technicians and the costs are subsidized by the State. It is sometimes criticized nowadays, usually for not working in situations for which it was never intended such as small-holder farming on steep land.

The components are shown in figure 10.6. Any runoff coming from higher ground above the arable land is diverted away from the arable by a

Plate 10.8 Tied-ridges effectively hold the rain until it can infiltrate

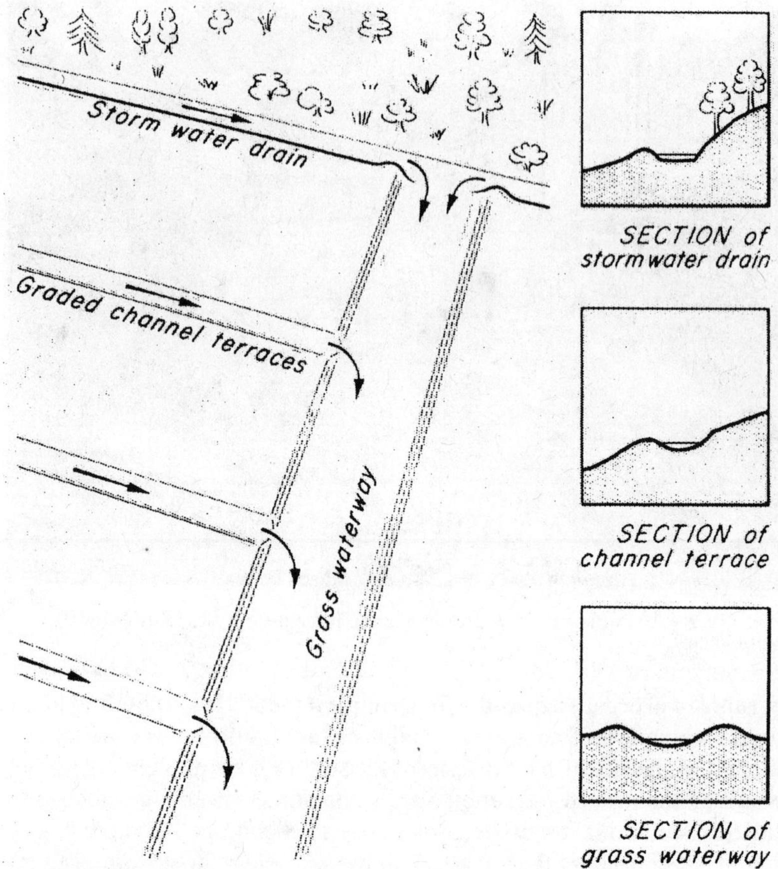

Figure 10.6 The basic components of mechanical protection:
(*a*) The stormwater drain which diverts storm runoff originating off the arable land
(*b*) The graded channel terrace leading away the runoff from the arable land
(*c*) The grassed waterway into which both stormwater drain and channel terraces discharge

cut-off drain, also called a *diversion drain* or *stormwater drain*. The runoff from the arable land is led off the field by *graded channel terraces* also called *contour bunds*, or in the USA just *terraces* (plate 10.10). The stormwater drain and the graded channel terraces discharge the runoff into a watercourse. Frequently there is no suitable natural watercourse and one must be artificially made. These are known variously as *grassed*

Plate 10.10 Graded channel terraces in Ethiopia

waterways, *sod waterways* or *meadow strips*, and will normally run straight down the slope with a bank on either side to contain the water. Each of these three channels must be carefully designed so that they can lead the surplus runoff safely away without silting or scouring of the channels, and they must ultimately discharge into a stream which can accommodate the additional flow. The channels in this system must be properly designed, carefully built, and regularly maintained, because if there is a failure of any of the channels the concentrated flow can cause serious erosion. An example is seen in plate 10.11 where a breach in the bank of a graded channel terrace at the top of the field starts a knock-on effect down the slope, with scour at all the consequent points of failure. The design procedures are now well-established and described in many text books and engineering manuals, a selection of which are listed in *Further reading*.

10.4.3 Controlled surface drainage

Surface open drains

Some tropical soils are surprisingly resistant to scour erosion in open channels and it is relatively easy to discharge surface runoff at unusually high velocities. An example is found in Ethiopia where the farmers traditionally have small open furrows to improve the surface

Plate 10.11 An example of the failure of channel terraces (Aerofilms Ltd)

drainage when growing teff (*Eragrostis abyssinica*), the staple cereal which has a tiny seed which is easily washed off the saturated surface. The furrows appear alarmingly steep at up to 15%, but they do not turn into gullies as might be expected, and are ploughed out and redrawn every year. The farmers are so convinced of the effectiveness of this system that they insist on cutting these steep channels through recently installed conventional channel terraces (plate 10.12).

Another example of surface drainage to solve a particular problem comes from Portugal where the situation is temporary waterlogging at the time of preparing for planting winter cereals. The method is called *pnudivales* which are small channels constructed after seed-bed preparation and before seeding. They are made with three passes of a tractor; first with a mouldboard plough to create a furrow, followed by two passes with a mounted ridger to produce a shallow open drain between 3 and 4

Plate 10.12 The traditional steep open drainage channels are often cut through recently constructed channel terraces in Ethiopia

mm wide and 0.25 mm deep at the centre. This allows subsequent drilling, cultivating, and harvesting in straight lines ignoring the open channels which are set out on a gentle gradient of 0.6–0.8% (SIMS 1983).

For some crops it is desirable to maximize the amount of surface runoff but this can lead to increased soil loss. Examples are tobacco and yams, both grown on ridges, and if these are set straight up and down slope there is increased erosion in the furrows on all but gently sloping land. Many years ago research on tobacco showed that having the ridges on a gentle gradient will give as good drainage as from steeper gradients, but at a non-scouring velocity with little soil erosion. For the light sandy soils commonly used for tobacco a grade of 2% was recommended (HUDSON 1957).

There are several other well established systems of field surface drainage. In the United States of America *bedding* is common on slopes up to 1.5% and is based on arranging the method of ploughing so that plough dead furrows are left at intervals of from 10 to 30 metres leading to a collecting ditch at right angles. Variations are *parallel field drains* which have larger cross-sectional areas and wider spacings than the dead furrows of bedding, and *parallel open ditches* still larger and which cannot be crossed with farm machinery. Design criteria for these systems are given in SCHWAB *et al* (1993). A variation of bedding which can handle the higher volumes of runoff associated with tropical rainfall is *ridge and furrow*. This combines an element of erosion control with surface drainage. The ground is tilled into wide parallel ridges of the order of ten metres wide, with intervening furrows about half a metre deep. Successive ploughings in the same direction build up the ridges over the years. As with bedding, surface runoff moves across the ridge to the furrow and then down the furrow which is on a gradient of about 1 in 400. The method is particularly suitable for large areas of gently sloping land, which do not quite justify channel terraces, but need some controlled surface drainage (PHILLIPS 1963). A variation in East Africa used on heavy black clay soils is called the *camber-bed system* (ROBINSON *et al* 1955, BROOK and ROBINSON 1959).

Yet another variation is the *broad-bed and furrow system* which has been mainly developed at the International Crops Research Institute for the Semi-Arid Tropics (ICRISAT) in India (PATHNAK *et al* 1985). An important component of the system is an ox-drawn wheeled toolbar, which can be used with ridgers to form the raised bed and also later for carrying seeders or planters (plate 10.13). The tool carrier is used for the initial forming of the beds, the subsequent annual reshaping, and for all tillage, planting, and inter-row cultivation. The cost of even the simplest factory-made toolbar is beyond the resources of many peasant farmers,

Plate 10.13 An ox-drawn wheeled toolbar developed at ICRISAT, India

but a really cheap wooden ridger shown in plate 10.14 shows promise.

In some circumstances surface drainage may be reinforced by underground drainage. This is usually expensive and only practical where an underground system of tile drains is already in place. The main advantage is that graded channel terraces can be carried across minor depressions which are already in place running down the slope. Runoff collects temporarily in the depressions behind the ridge of the terrace and flows through vertical surface inlets connected to the tile drains.

10.5 Terracing and cross-slope barriers

The many forms of terracing may be divided into those where the whole result is completed in a single operation (sometimes called *one-shot terracing*), and those where the land slope is progressively changed in small increments over a number of years. The latter are discussed in section 10.6.

10.5.1 Conditions suitable for terracing

A prodigious amount of labour is required for the construction of bench terraces so, although they are commonly found in many parts of the world, it is more often as a legacy from the past than present day

Plate 10.15 The ancient bench terraces of Machu Picchu, in Peru (V. AUSTIN)

timber. In very stable soils the walls may be held only by vegetation. Variations of bench terracing are for the cultivation step to be horizontal, or to have a slight slope outwards (plate 10.16), or a slope backwards towards the hill, called *reverse-slope terraces* (figure 10.7). An example of ancient bench terraces in the Yemen Arab Republic is shown in plate 10.17.

Bench terraces may be constructed on the contour to minimize runoff, or with a slight gradient like graded channel terraces. The difficulty is to discharge any surface runoff flowing along a bench terrace down the slope without causing erosion. Land which is bench terraced is usually too steep for vegetated waterways to be effective, but channels lined with stone or concrete add greatly to the cost and maintenance requirements.

Irrigation terraces Sometimes the flat bench terrace has a raised lip at the outer edge to retain irrigation water, extensively used for production of rice, and to a lesser extent for tea, fruit trees, and other high-value crops. For paddy, the terraces are also level along their length, so that each becomes a flooded shallow pond, but where intermittent water application is intended there can be a slight longitudinal fall similar to that for border irrigation. The Banaue rice terraces in the Philippines are a

Plate 10.16 Outward-sloping bench terraces in Uttar Pradesh, India

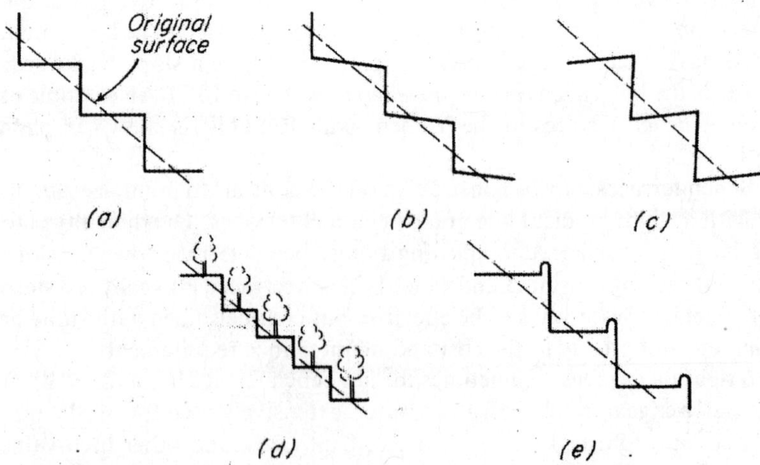

Figure 10.7 Types of bench terraces
(*a*) Level bench (*b*) Outward-sloping bench (*c*) Inward-sloping (or reverse-slope) bench (*d*) Step terraces (*e*) Irrigation terraces

Plate 10.17 Bench terraces in the Yemen Arab Republic (H. VOGEL)

magnificent example of irrigation terraces, a monumental engineering feat of great antiquity, but still maintained today (plate 10.3). This type of terrace is found in most regions where rice is a staple food, and is very common in Indonesia, Malaysia, China, and Japan. Irrigation terraces for wheat are shown in plate 10.18 from Bhutan.

The design of bench terraces The physical factors are the steepness of the land and the depth of the soil. The management decisions are the width of the terrace, the height of the riser, and the slope of the riser.

Slope The most important factor is the steepness of the land slope. Theoretically terraces can be built on any slope with deep soil, but in practice on very steep slopes the riser becomes too high and consequently difficult to maintain, and the terrace becomes too narrow. Practical limits are about 20° for terraces built by machine, and 25° when built by hand. Intermittent terraces (section 10.5.3) can be built up to 30°. It is not usual to use bench terraces on slopes less than 7° because control can be more economically achieved through graded channel terraces. The steepness also affects the amount of land which is available for cultivation after terracing. On steep slopes the land taken up by the sloping riser and possibly toe drains can be up to a quarter or a third of the area available without terracing, and where there is a serious shortage of available arable land, this can be powerful deterrent.

Plate 10.18 Level irrigation terraces in Bhutan

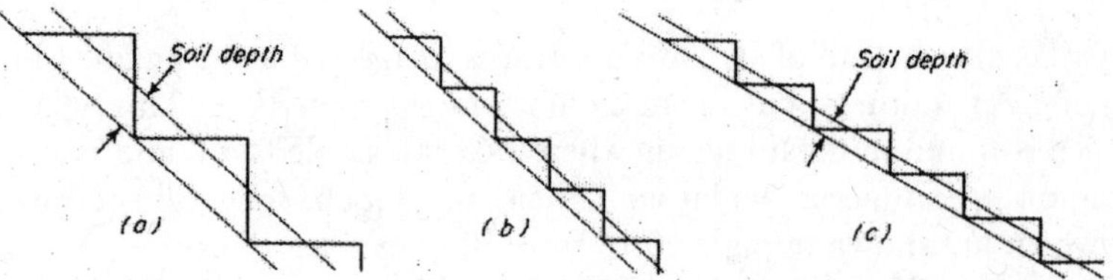

Figure 10.8 For a given slope, a greater depth of soil allows wider terraces (*a*, *b*)
For a given soil depth, a gentler slope allows wider terraces (*b*, *c*)

Width The width of bench terraces is determined by three factors: the cost, what is desirable for ease of cultivation, and what is practical for construction. For a given slope increasing the width of terrace increases the amount of excavation and the amount of fill. Narrow terraces are therefore cheaper than wide terraces. For terraces cultivated by hand or walking tractor, bench widths of from two to five metres are suitable. For animal draft or four-wheel tractor cultivation, a minimum width of three or four metres is desirable.

The maximum width which is economically practical is determined by the combination of land slope and soil depth as shown in figure 10.8. It is not usually practical to excavate down into the sub-soil or bedrock, both because of the increased cost of excavation and the lower fertility of the exposed sub-soil on the inner edge of the terrace. This may be alleviated by careful placing of the soil, i.e. building one terrace and then putting on it topsoil removed from the next terrace, as shown in figure 10.9.

The earth-moving required is also substantially reduced if the terrace is cut with an outward slope instead of a horizontal bench as shown in figure 10.10. This is normal practice in many countries along the Himalaya range from Kashmir to Bhutan as illustrated in plate 10.16. Outward sloping bench terraces are more likely to spill over the terrace edge than level terraces, and this can cause problems on erodible soils. Grass planted on the terrace edges will help, and there will be an element of progressive levelling of the terrace as discussed in section 10.6.

When regular cultivation is not required, as for tree crops, smaller terraces will be equally effective and will need less earth moving. These are usually called *step-terraces*, and used for fruit trees, tea, coffee, and vines (figure 10.1 and plate 10.19). Manuals on the design and construction of terraces and ditches are listed in *Further reading*.

10.5.3 Intermittent terraces

This is a collective term sometimes used for terraces or drains cut at intervals down the slope, while retaining the original land slope between terraces. They are usually used when growing tree crops such as spices or rubber, as in plate 10.20. The terrace serves as a path for collecting the harvested latex, and it is also useful to have a flat area for weeding round each tree, and for applying fertilizer. They are also used for fruit trees, hence the alternative name *orchard terrace*. If harvesting paths are not required the terraces need not be continuous, for example in oil palm plantations where they are described as *platforms* (plates 10.21 and 10.22). The important feature of any of these development techniques for steep erosion-prone slopes is that the land between the terraces must be planted to a vigorous cover crop. Creeping legumes such as *pueraria*, *centrosema*, and *calopogonium* are suitable. Another practice used on steep slopes is *hillside ditches* (plate 10.23). The term is applied to several shapes, some of which are shown in figures 10.11 and 10.12, the common feature being a ditch dug on the contour to catch soil and water. A variation is the *lock and spill drain* (plate 10.24 and figure 10.12). The ditch is about 0.5 m^2 on a slight grade. Low cross walls are left in the bed of the drain so that it is divided into separate basins (*locks*) and so encourage infiltration. In heavy rain, runoff overtops these cross walls and *spills* towards the outlet, often a stone-lined channel. Periodically the accumulated soil is emptied and spread on the land between the ditches.

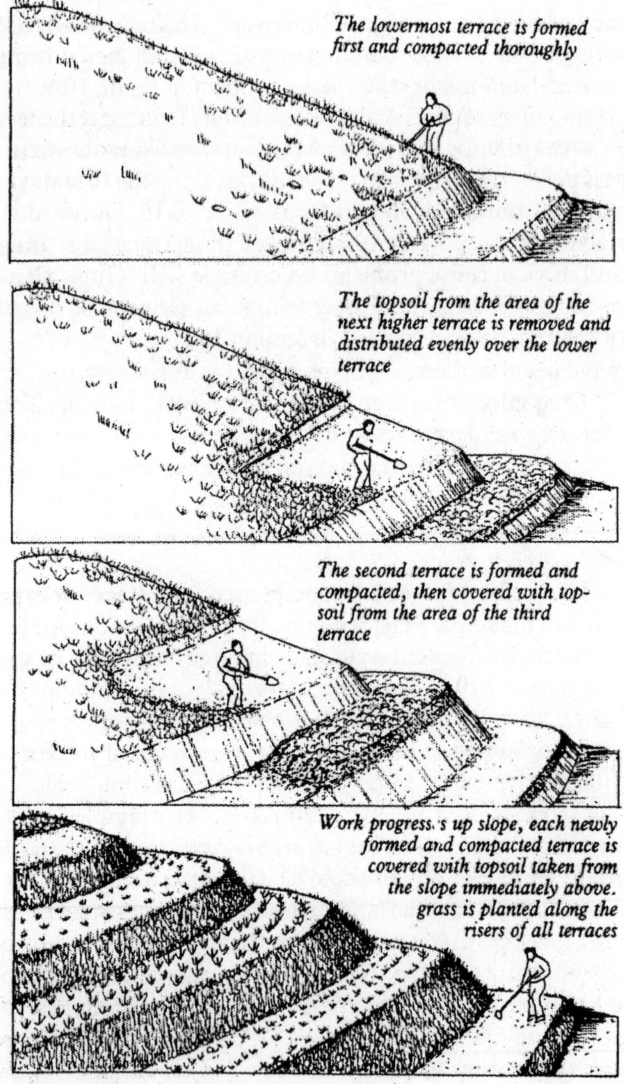

Figure 10.9 Making bench terraces without burying topsoil (PEACE CORPS 1986)

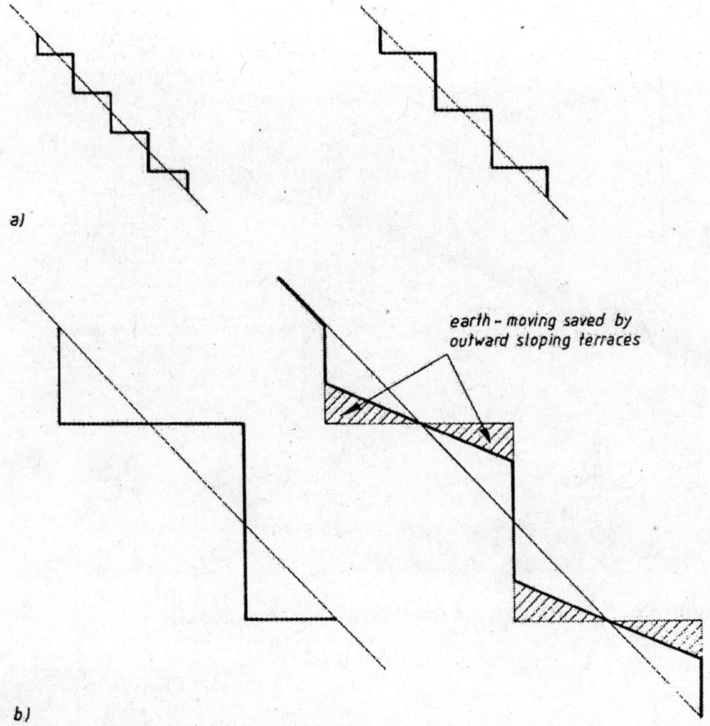

Figure 10.10 Earth moving required for terracing
(*a*) Wider terraces mean more earth moving. The volume is proportional to $\left(\frac{\text{width}}{2}\right)^2$
(*b*) Outward-sloping terraces mean less earth-moving

10.6 Progressive terracing

Progressive terracing can sometimes be a sensible alternative to terracing in a single operation. When labour is in short supply it can make it possible to spread the effort of earth moving over a number of years, and it also allows the terracing effect to be assisted by cultivation. The amount of soil moved across the terrace by erosion is sometimes quoted as the main factor in the levelling process, but in fact the movement as a result of cultivation is the more important process. Studies many years ago measured the change in profile as a result of different methods of ploughing, and showed that there is not much movement upslope even as

Plate 10.19 Step terraces are small bench terraces (USDA)

Plate 10.20 Intermittent terraces planted to rubber in southern Thailand. The slope between the terraces must be planted with a crop cover

Plate 10.21 Platforms are small sections of terrace for individual plantings

a result of ploughing techniques intended to encourage this, but there is considerable movement downslope (HUDSON 1964).

10.6.1 Progressive terracing on arable land

This method is traditional in most of the countries bordering the Himalaya mountain range (NEPALI 1981), and it was included in the official recommendations of the USDA-SCS for Puerto Rico as long ago as 1941 (figure 10.13). That method and the recent variation used in Venezuela (WILLIAMS and WALTER 1988) both used stone walls which are progressively raised as the terrace builds up. This requires much labour, but may be an acceptable solution if suitable stone is readily available in the fields, and using it for terracing will make cultivation easier. Plate 10.25 shows progressive terracing using simple walls of stone collected from the ground between the terraces.

Where stone is not available locally, progressive terracing can be developed using earth banks. A very successful system comes from Kenya

Plate 10.22 Platforms for oil palm in Malaysia

Plate 10.23 Reverse slope terrace, or hillside ditch in El Salvador

Figure 10.11 Hillside ditch, individual platform terraces, and orchard terraces (from PEACE CORPS reprint R-62)

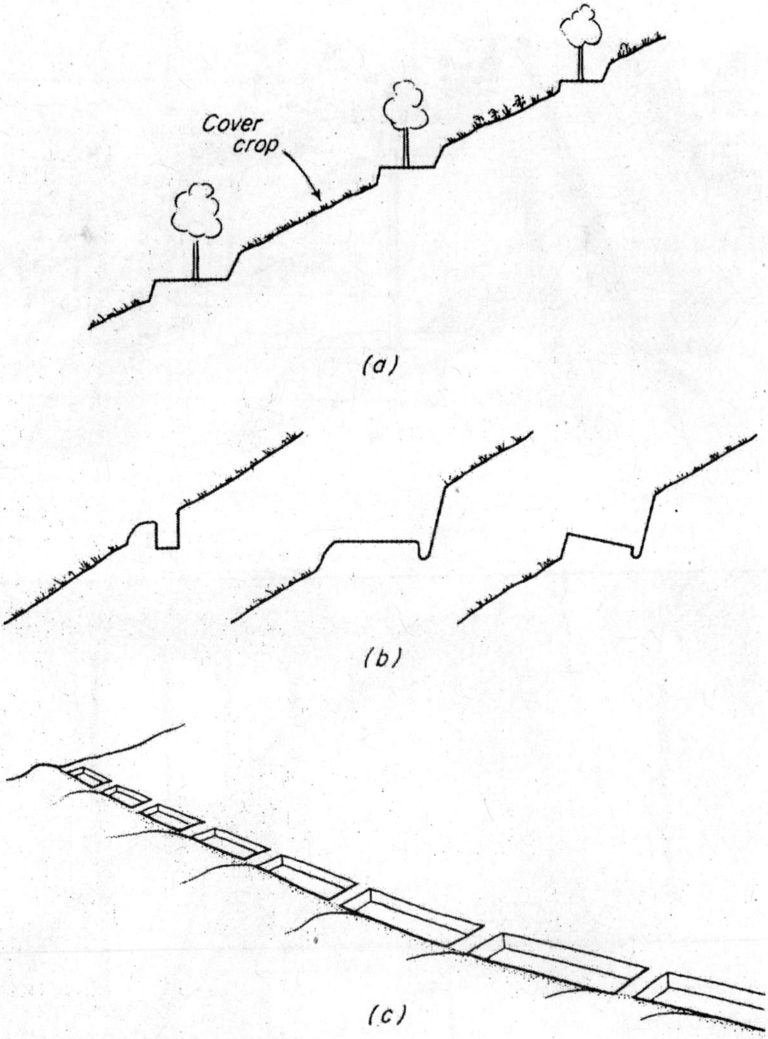

Cover crop

(a)

(b)

(c)

Figure 10.12 Intermittent terraces
(*a*) Orchard terraces
(*b*) Types of hillside ditch
(*c*) Lock and spill hillside ditches
(*d*) Fanya juu terraces (Kenya)

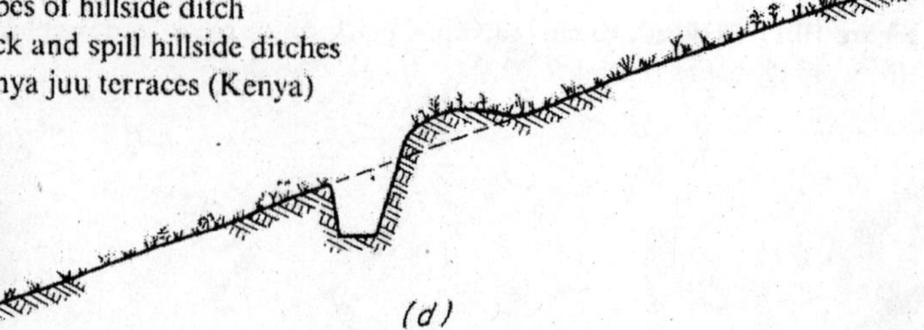

Plate 10.24 Hillside ditch with lock and spill drain (USDA)

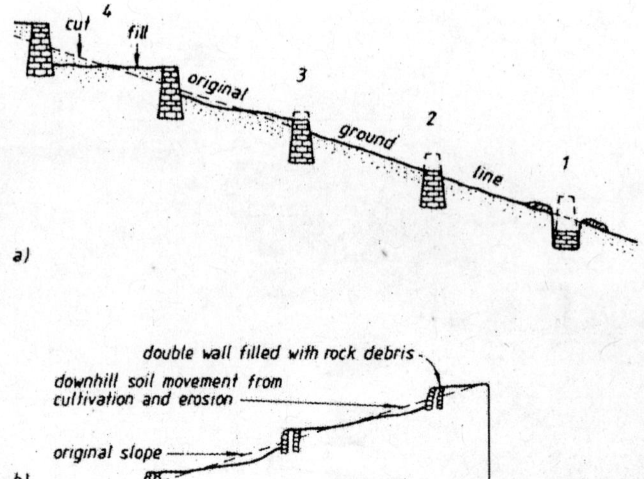

Figure 10.13 Examples of progressive formation of bench terraces
(*a*) Puerto Rico (USDA–SCS 1941)
1 Trench excavation made
2 Initial masonry completed and trench back filled
3 Bench starts to form and wall raised
4 Completed bench terrace
(*b*) Venezuela (WILLIAMS and WALTER 1987)

and is known as *fanya juu* which means to throw upwards, as opposed to *fanya chini* to throw downhill as in the construction of channel terraces (THOMAS *et al* 1980). The principle of fanya juu is to build a bank by excavating a ditch and throwing the soil on the uphill side (plate 10.26 and figure 10.12). Soil starts to build up above the bank as it is moved downslope by cultivation and erosion. This may be accelerated by planting grass on the ridge (plate 10.27). In subsequent years the bank is built higher by throwing up more soil from the terrace below. The labour input can be further spread over time by first building the terrace at a wide spacing, then later building intermediate terraces as in figure 10.14.

This method is very suitable in central Kenya where deep red-clay loams have a high moisture storage capacity, and with a good vegetation cover most of the rain will infiltrate. If runoff does occur, there may be some risk of damage as if flows over the terrace bank, but this can be contained by planting vigorous grasses such as weeping love grass (*Eragrostis curvula*), or makarikari (*Panicum coloratum makarikariensis*)

Plate 10.25 Stone walls built from surface stone make cultivation easier and have a terracing effect (WORLD NEIGHBORS)

Plate 10.26 The fanya juu type of intermittent terrace used in Kenya

Plate 10.27 Vegetation planted on the ridge of fanya juu terraces increases the deposition of silt

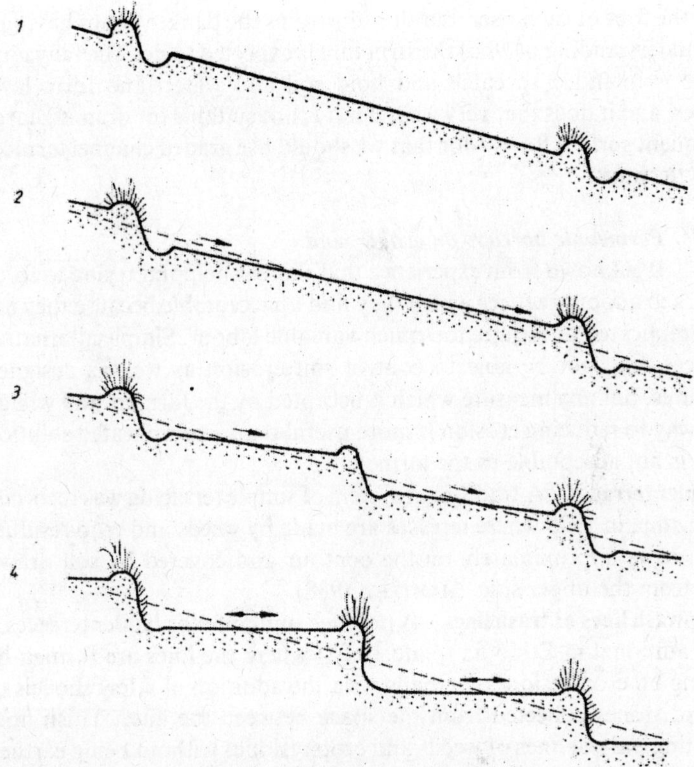

Figure 10.14 The fanya juu method of terracing
1 Terrace is built at 2 m vertical interval by throwing excavated soil up-hill to form a ridge, which is planted with fodder grasses
2 Movement downslope by cultivation and erosion starts to build up a lower terrace
3 Main terrace banks built up higher, and intermediate terrace added
4 Final profile is nearly level terraces with well-vegetated terrace risers

on the bank, and used for cut-and-carry cattle feeding (plate 10.27). Fruit trees are often grown on the edge of the terrace, so there is no loss of production as a result of some land being taken out of cultivation. Like most successful ideas, fanya juu terracing is sometimes over-enthusiastically welcomed and used in unsuitable conditions, particularly where large runoff flows can be regularly expected. Some attempts have been made to cope with this situation by using fanya juu terraces on grade, and hoping that the water will be drained away by the small drains sometimes left at the foot of each riser, but this illustrates the danger of not having a clear understanding of what the structure is expected to do. The fanya juu terrace is intended to catch and hold soil and water, and form level terraces, and it does that very well, but it is not suitable for draining large or frequent surface flows – for that we should use graded channel terraces and waterways.

10.6.2 Permeable barriers on arable land

It is known from experience that there is no point trying to force farmers to adopt practices which they find unacceptable because they are too complicated or require too much valuable labour. Simple alternative practices may not be able to control soil erosion as well as designed structures, but any measure which is accepted by the farmers and will go some way to reducing erosion is more useful than a complicated solution which is not acceptable to the farmers.

Ladder terraces A traditional system of simple terracing was recorded in Tanzania in 1938 where terraces are made by weeds and crop residues laid in rows approximately on the contour and covered by soil drawn down from the upper side. (HARTLEY 1938).

Stopwash lines or trash lines A practice similar to the ladder terraces is also traditional in Ethiopia (plate 10.28) where the lines are formed by building up crop residues, perhaps with the addition of a few shovels of soil and stones collected from the space between the lines. Trash lines consisting only of lines of weeds and crop residues without being earthed over are also commonly found throughout east and southern Africa. All of these simple methods will be less effective on very steep slopes, or where they may be washed away by heavy runoff.

Grass strips Grass strips can also be used as stop-wash lines, and this was the basis of a national conservation programme in Swaziland. In the 1940s the King issued a royal edict that strips of the indigenous grass were to be left on all ploughed land, two metres wide at two metre vertical intervals. The rule was rigorously enforced and almost all arable land has grass strips today as shown in plate 10.29. For lack of sufficient field advisors many of the strips were off-grade, and others were on land which is too steep for erosion to be halted by this method, as shown in plate 10.30, but erosion in Swaziland would be very much worse if these strips had not been left. In Kenya live hedges are sometimes planted for livestock control, often using sisal, euphorbia, or other drought-resisting species (plate 10.31). In areas with reliable rainfall, high yielding grasses may be densely planted to cut for fodder as in plate 10.32 (from Brazil). The choice of grass should be based on several factors; what is known to

Plate 10.28 Simple stop-wash lines of straw, stones, and soil in Konso, Ethiopia (H. HURNI)

Plate 10.29 Grass strips have been in place in most arable land in Swaziland for the last 40 years

Plate 10.30 Grass strips reduce but do not entirely prevent erosion on steep land in Swaziland

grow vigorously in the district, trials of local indigenous species and imported varieties should always be carried out, and the use of the grass should be considered. A palatable grass of high feed value would be indicated where it is to be cut and carried for stall feeding, but where cattle have free access to arable land after harvesting it may be more appropriate to use an unpalatable grass to avoid the cattle trampling the banks. A major campaign is currently being undertaken by the World Bank for the use of vetiver grass (*Vetiveria zizanioides*) (US NATIONAL RESEARCH COUNCIL 1993). There is no doubt that vetiver has been successfully used in many countries, but there is a danger in pushing its adoption too hard because no grass is a universal magical solution and all the possibilities should be considered and tested locally. In Kenya, KIEPE and YOUNG (1992) compared the terracing effect of fanya juu terraces, contour hedges of *Leucaena leucocephala* and *Grevillia*, and other trees planted on grass strips of *Panicum coloratum*. After six years all three practices showed a significant levelling of the cropped land, with the average original slope of 14% halved to about 7%. The much lower labour requirement for planting contour hedges or grass strips makes them practical alternatives to earth-moving practices.

Plate 10.31 Live hedges of drought-resisting species such as sisal and euphorbia are used to contain livestock in Kenya

The use of grass strips and other permeable barriers on grazing land is discussed in section 13.1.3.

References

ELWELL, H. A. and A. J. NORTON 1988 *No-till tied-ridging: a recommended sustained crop production system*, Institute of Agricultural Engineering, Harare, Zimbabwe

FINKEL, H. J. 1986 *Semi-arid Soil and Water Conservation*, CRS Press, Boca Raton, Florida

HARTLEY, B. J. 1938 An indigenous system of soil protection, *East African Agriculture Journal*, 4, 63–66

HUDSON, N. W. 1957 Soil Erosion and Tobacco Growing, *Rhodesia Agricultural Journal*, 54, 6, 547–555

HUDSON, N. W. 1964 The effect of ploughing methods on the slope of the land between ridges, *Rhodesia Agricultural Journal*, *Bulletin 2247*, May/June 1964

HULUGALLE, N. R. 1988 Properties of tied ridges in the Sudan savannah of the west African semi-arid tropics, in *Land Conservation for Future Generations*, edited by S. Rimwanich, Department of Land Development, Ministry of Agriculture, Thailand

Plate 10.32 Grass strips used to create a terracing effect and to provide fodder in Brazil (H. LAL)

KIEPE, P. and A. YOUNG 1992 Soil conservation through agroforestry: experience from five years of demonstrations at Machakos, Kenya, chapter 24, in *Erosion, Conservation, and Small-scale Farming*, edited by H. Hurni and K. Tato, Geographica Bernensia, Berne, Switzerland

NEPALI, S. B. 1981 Soil erosion and traditional know-how, in *Problems of Soil Erosion and Sedimentation*, 371, edited by Tingsanchali and Eggers, Proceedings of South East-Asian Regional Symposium at Asian Institute of Technology, January 1981

NYAMUDEZA, P., E. MAZHANGARA, T. BUSANGAVANYE and E. JONES 1993 Farmer adoption of improved water management on vertisols in semi-arid south-east Zimbabwe, in *Working with Farmers for Better Land Husbandry*, edited by N. Hudson and R. J. Cheatle, Intermediate Technology Publications, London

PATHAK, P., S. M. MIRANDA and S. A. EL-SWAIFY 1985 Improved rainfed farming for semi-arid tropics: implications for soil and water conservation, in *Soil Erosion and Conservation*, 338–354, edited by S. A. El-Swaify, W. C. Moldenhauer and A. Lo, Soil Conservation Society of America, Ankeny, Iowa

PEACE CORPS 1986 *Soil Conservation Techniques for Hillside Farmers*, Peace Corps, Information Collection and Exchange, Washington DC

PHILLIPS, R. L. 1963 Surface Drainage Systems for Farm Lands (Eastern United States and Canada), *Transactions of the American Society of Agricultural Engineers*, 6, 313–317, 319

ROBINSON, J. B. D., T. R. BROOK and H. H. DE VINK 1955 A Cultivation System for Ground water (Vlei) Soils – Part I, *East African Agricultural Journal* 21, 69 (Part II – Modification of the System, by T. R. Brook and J. B. D. Robinson, 25, 192, 1959)

SCHWAB, G., W. ELLIOTT, D. FANGMEIER and R. K. FREVERT 1993 *Soil and Water Conservation Engineering*, 4th edition, Wiley, New York

SIMS, D. 1983 *Pnudivales, a practical and effective way of reducing waterlogging and preventing soil erosion on cereal land*, Project Working Paper, Drainage and Soil Conservation in the Alentejo Region of Portugal POR/81/002, FAO, Rome

SMITH, P. D. and W. R. S. CRITCHLEY 1983 The potential of runoff harvesting for crop production and range rehabilitation in semi-arid Baringo, in *Soil and Water Conservation in Kenya*, Proceedings of Second National Workshop, Nairobi, Kenya, March 1982, Occasional Paper 42, Institute of Development Studies, University of Nairobi

SWAMINATHAN, M. S. 1982 *Rainfall and Dryland Agriculture*, Proceedings of Symposium on Rainwater and Dryland Agriculture, IARC, New Delhi, October 1980

THOMAS, D. B., R. G. BARBER and T. R. MOORE 1980 Terracing of Cropland in Low Rainfall Areas of Machakos District, Kenya, *Journal of Agricultural Engineering Research*, 25, 1, 57–63

USDA-SCS 1941 *Soil Conservation in Puerto Rico and the Virgin Islands of the United States*, USDA-SCS, San Juan, Puerto Rico

US NATIONAL RESEARCH COUNCIL 1993 *Vetiver grass: A thin green line against erosion*, Board on Science and Technology for International

Development, National Academy Press, Washington DC
WILLIAMS, L. S. and R. J. WALTER 1988 Controlled-erosion terraces in Venezuela, chapter 19, in *Conservation Farming on Steep Lands*, edited by W. C. Moldenhauer and N. W. Hudson, Soil and Water Conservation Society, Ankeny, Iowa

11

Erosion control through land husbandry

11.1 The farmer's point of view

In the last chapter possible conservation practices were reviewed in groups according to their purpose in terms of influencing runoff or soil loss. Now, looking at ways of limiting soil degradation by improved land husbandry, the primary consideration is to look at the alternatives from the farmer's point of view. In fact we need to be very wary of manual-type recommendations. A better approach is to offer a menu of alternatives, and the role of the professional is to help the farmer choose what is appropriate for his or her circumstances. A recent review of technologies acceptable to resource-poor farmers points out that combining practices into a farming system must take account not only of the physical factors such as soil and climate, but equally the available resource inputs, especially cash and labour, and the farmer's objectives (STOCKING 1993).

The objectives of a commercial farmer are usually to maximize yield and income, but the subsistence farmer is likely to be more interested in improving food security by reducing the risk of failure, or improving the return on inputs of seed, fertilizer and labour, or improving the quality of life by reducing drudgery, particularly relevant in the case of women farmers.

We need a new approach to the farmer when recommending changes to farming practice. Instead of recommending practices because they will reduce erosion, we should show how they can lead to increased production. For example, practices which lead to better soil structure, more organic matter, or more moisture-holding capacity should be put forward because they improve yield, not because they reduce the soil erodibility. Increasing infiltration from better ground cover or the use of contour farming or ridging can result in better crops as well as reducing soil loss. Improved crop management could include more use of fertilizer and manures or crop rotations. These practices will make more sense to the farmer if they are recommended as leading to better production, which he can instinctively appreciate as desirable, rather than because they will reduce soil erosion which is likely to be a less important objective.

11.2 Erosion control by good farming

It is remarkable how often the management required for good erosion control coincides with intensive, efficient, profitable farming. There is no substance at all in the fear sometimes felt by farmers that conservation farming means restricted production or uneconomic practices.

This link between reduced erosion and improved farming can best be demonstrated by some examples. The first is the case of growing tobacco in the sub-tropics. The growing season for flue-cured tobacco is shorter than the rainy season, and the alternatives open to the grower are either to plant early and have the crop matured and harvested well before the end of the rains, or to delay planting so that the end of the harvesting coincides with the end of the rains. In both cases there will be a period either at the beginning or at the end when there is poor vegetation cover and so high risk of erosion. Field experiments showed a soil loss of 4.7 metric tons per hectare from the early planted crop compared with 8.7 metric tons per hectare from the late planted crop (HUDSON 1957a). It was also found by the Tobacco Research Board that the yield and quality of the tobacco are better from the early planted crop, and so there was a direct link. The same practice was desirable both for erosion control and better crop management.

A similar situation exists in the case of the management of grazing land. This is discussed more fully in section 13.1.2 but the main point is that in order to achieve maximum yield, whether of meat, wool, milk or any animal products, certain conditions are required: correct stocking rate, uniform distribution of grazing, optimum grass growth, and so on. Again these are precisely the things which are required for optimum soil conservation, and so good management equates with good erosion control.

In the 1970s it was topical to classify crops as *soil-depleting* and *soil-conserving*, and to speak of *soil robbers* or *soil builders*. Fashions change slowly in agriculture and this concept still crops up occasionally today although it has little relevance in modern agriculture. The fact is that it is not the crop which is or is not soil depleting, but the crop management. Loss of fertility is not so much a question of *which* crop is

Table 11.1 The effect of crop management on the soil and water losses from maize

Plot A *Maize at medium level of production*		*Plot B* *Maize at medium level of production*
25 000 plants/ha	Plant population	37 000 plants ha
N 20 kg/ha	Fertilizer application	N 100 kg/ha
P_2O_5 50 kg/ha		P_2O_5 80 kg/ha
Removed	Crop residues	Ploughed in
5 ton/ha	Crop yield	10 ton/ha
250 mm	Runoff	20 mm
12.3 ton/ha	Soil Loss	0.7 ton/ha

Results for season 1954/55 (rainfall for the season 1130 mm)
From HUDSON 1957b

grown but *how* it is grown. Certainly some crops will be particularly prone to erosion because of the way in which they are grown, particularly clean-tilled row crops. Other closely-growing crops such as grass will usually suffer less erosion. But these general trends can be overruled by the effect of management. Grass, if overgrazed and under-fertilized, can allow so much erosion that instead of improving the soil the fertility actually declines. And maize can under suitable management both reduce erosion and improve the soil, as was demonstrated over 30 years ago in Zimbabwe. Soil and water losses were measured from plots where maize was grown continuously both according to current practice, and also according to the practice recommended by the advisory services for maximum yield. The crop treatments and soil and water losses are shown in table 11.1. Both the soil and water losses were reduced to something like one fifteenth by the improved crop management. The high yielding crop produced twice as much grain as the lower level of production, and the fertility and the yield were maintained for ten years while the production at the lower level steadily declined. In fact the effect of applying the maximum-yield crop management treatment for many years completely disproved the fallacy that continuous maize is necessarily a bad practice. In this case after ten years the soil was in better condition physically than at the start of the experiment, the moisture status better, the yield maintained, and the soil loss during the whole period was negligible.

11.3 Conservation farming

This title covers many different farming techniques. It includes any farming practice which improves yield, or reliability, or decreases the inputs of labour or fertilizer, or anything else leading towards improved land husbandry. Sometimes there is a long history of traditional farming and soil conservation practices which have been tested and developed over periods of time which are long enough to include all the likely variations of climate. These traditional practices should give the best long-term result, bearing in mind that the farmer's interpretation of 'best' may be based on reliability rather than maximum yield. Possible new techniques should have the same basic characteristics as traditional practices, that is they should be easy to understand, simple to apply, have low inputs of labour and capital, and must show a high success rate, i.e. a high rate of return on inputs. Some of the techniques are:

- *Farming on a grade* is well established in India (SWAMINATHAN 1982). Cultivations and planting are done on a gentle gradient which encourages infiltration but permits surplus runoff at low velocities. Sometimes this may be combined with simple practices to encourage infiltration such as returning the crop residues. There can still be a problem of disposal of the surface runoff when it does occur.
- *Strip cropping* is most useful on gentle slopes, where it may reduce runoff to acceptable levels without any banks or drains.
- *Rotations* are another well-established and simple practice. The object may be to improve fertility by the introduction of legumes into the rotation, or to improve structure by the introduction of periods of grass, or to help control the progressive build-up of pest or disease. Crops with different root patterns can vary the uptake of moisture or nutrients; *nurse crops* can prepare the way for a subsequent crop; rotations in space rather than in time can give some protection against climatic variation – there are many reasons why crop rotations are so widely used.
- *Shifting cultivation or bush fallow* has been widely practised for centuries in all continents, particularly on land too steep for continuous cultivation, and like every other form of land use it can be damaging or sustainable according to how it is carried out. With small areas of ground cleared for crops and long fallow periods between cropping, the system is efficient in terms of labour use, and unless the erosion during the cropping period is particularly severe, the long-term average soil loss need not be excessive. Plate 11.1 shows an example from Venezuela. The problem is that if pressure on the land leads to large areas being cleared and short fallow periods between the cropping, the land

Plate 11.1 Shifting cultivation on steep forest in Venezuela (I. PLA SENTIS)

does not recover during the fallow, and the system is not sustainable. The usual practice is to leave the land to regenerate naturally, but the fallow period can be shortened by sowing or planting grasses or legumes.

- *Fallowing* is well established as part of rotations, particularly in regions of low rainfall, but is not always successful and must be tested in local conditions. A season of cropping may be alternated with a season of bare fallow, or winter fallow followed by a summer crop or vice versa. The advantages claimed are conservation of soil moisture during the fallow, and accumulation of nitrates in the soil surface. In the drier wheatlands of Australia a bare fallow in summer is used to build up soil moisture before sowing the winter wheat which receives only barely adequate rainfall. A similar practice is followed in India where a bare fallow with frequent cultivation is started towards the end of the rainy season (*kharif*) and followed by a crop of pulse or cotton grown during the dry season (*rabi*) on accumulated moisture. Other examples can be found in FAO *Soils Bulletin* 57 (1987).

- *Mixed cropping and interplanting* are widely applied traditional techniques. A combination of crops with different planting times and different length of growing periods spreads the labour requirement of planting and of harvesting, and also allows mid-season change of plan according to the rain in the early part of the season (SWAMINATHAN 1982, STEWART 1988). Another possible advantage may arise from the use of legumes to improve the nitrogen status for the cereal crop.
- *Timeliness* of farming operations is always important particularly where the rainfall is erratic. Yields can be dramatically affected by planting or cultivating at the right time, but common problems are having to wait for rain to soften the ground because it is too hard to plough when dry, and perhaps then not being able to plant because the ground is too wet, or a family with only one ox having to wait to borrow another one, or having to wait for a month after the rain starts to get the oxen back into condition for ploughing after a hard dry season.
- *Deep tillage* One of the reasons for low yields in semi-arid areas is the limited amount of moisture available to crop roots. This may be increased if the rooting depth is increased, and it has been shown that in some cases deep tillage can help, for example in the dense sandy soils (luvisols) in Botswana (WILLCOCKS 1984). Deep tillage requires greater draft power, which is usually in short supply in semi-arid areas.
- *Ripping* or *subsoiling* can be beneficial, either to increase the porosity of the soil, or to break a pan which is reducing permeability. The deep placement of fertilizer can also be used to encourage more rooting at depth, but the application of this technique to subsistence farming will be difficult.

11.4 The importance of cover

The dramatic reduction in erosion and runoff from a well-covered soil compared with a bare soil was clearly demonstrated in an experiment in Zimbabwe 40 years ago. Two small experimental plots,

Table 11.2 Soil losses from bare and covered soil

	Plot covered by gauze	*Bare plot*
1953/54	Nil	146.2 ton/hectare
1954/55	2.0 ton/hectare	204.5 ton/hectare
1955/56	4.5 ton/hectare	135.6 ton/hectare
1956/57	0.2 ton/hectare	132.4 ton/hectare
1957/58	0.2 ton/hectare	49.5 ton/hectare
1958/59	2.5 ton/hectare	202.0 ton/hectare
1959/60	Nil	7.4 ton/hectare
1960/61	Nil	121.4 ton/hectare
1961/62	Nil	138.5 ton/hectare
1962/63	Nil	128.2 ton/hectare
10 year totals	9.4 ton/hectare	1265.7 ton/hectare

From HUDSON 1957b

shown in plate 4.4, were 1.5 metres wide by 27.5 metres long. Both were kept free of vegetation by hand weeding, and over one plot was suspended a double layer of fine-mesh wire gauze. This allowed all the rain to pass through (this was measured by rain gauges on each plot, but broke up the falling raindrops so that they reached the soil surface as small droplets). The two plots thus compared the effect of no protective cover with a 100% cover, and the resulting soil loss is shown in table 11.2. The soil loss from the bare plot is more than one hundred times than that from the protected plot.

A vegetative cover as complete as the artificial cover of the wire gauze is only obtained from grass and forage crops. In the case of row crops only part of the soil is covered, and in the case of broadcast crops such as small grains, the cover may be uniform but is not dense, and so in both cases the effect of variations in the amount of cover is important. It is evident that small differences in cover can cause big differences in soil loss. Table 11.1 gave the results of a comparison of maize grown at two levels of production and the erosion differed 15 times. If differences of this size can occur with different densities of the same crop, then it is not surprising that there can be even bigger variations between different kinds of crop.

The reason why small changes of density of cover cause big changes in erosion becomes clear when the erosion process is analysed. Erosion takes place from the soil which is damaged by the rain, so the amount of erosion depends on how much of the soil is exposed, not how much is covered. The point is best illustrated by example. Photographs were taken vertically downwards on various densities of crops at various stages of growth and the percentage of the ground covered by the crop was measured on the

photographs. Plate 11.2 shows the maize on the two plots of table 11.1, one with 24 000 plants per hectare, the other with 36 000 plants per hectare, both being planted at the same time and photographed at the same time after six weeks growth. Calculations on similar photographs showed that at 10 weeks the lower population covered 60% of the ground leaving 40% exposed. The higher population covers 90% of the ground, leaving 10% exposed. The cover is only increased by 50%, but the erosion is proportional to the bare ground and is reduced to one quarter. This was in fact precisely the long-term result of the experiment referred to in section 11.2, table 11.1, when erosion was measured from maize grown at these two populations. In individual years the difference was as much as 15 times, as quoted in table 11.1, but the average over ten years was that the high population plot lost one quarter of the soil lost from the lower population, corresponding precisely with the relative amounts of exposed soil.

Plate 11.2 Photographs taken vertically downwards show the difference in the amount of soil covered by maize at 2 400 plants per hectare (left) and 3 600 plants per hectare (right)

A large reduction in erosion can thus result from quite small changes in management practice. This argument should not be carried to extremes; it does not follow that a grass crop with 98% cover and 2% exposure will have twice the erosion of a crop with 99% cover and 1% exposure, for erosion would be insignificant in both cases with cover so dense. But in the range of cover which is most usual for arable crops, variations in the density of cover will influence erosion more than any other management factor. The general relationship between splash erosion and cover is illustrated in figure 11.1, and has been confirmed by experiments from every continent as reported in the proceedings of ISCO conferences 3, 4, 5, and 6 (see *Further reading*).

11.5 Conservation tillage

This is another umbrella term and can include reduced tillage, minimum tillage, no-till, direct drill, mulch tillage, stubble mulch farming,

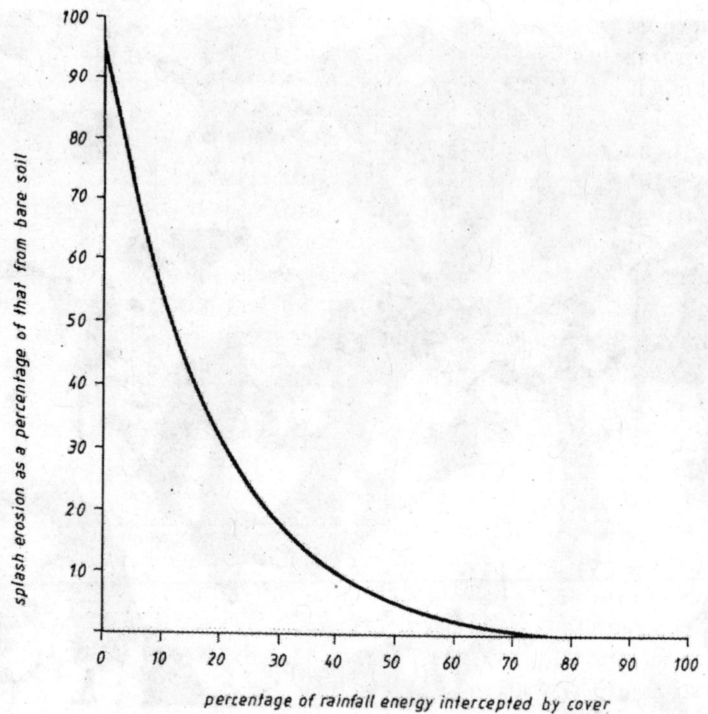

Figure 11.1 The effect of ground cover in reducing splash erosion (SHAXSON *et al* 1989)

trash farming, strip tillage, and plough-plant (MANNERING and FENSTER 1983). In the USA and Australia the concept of conservation tillage is a main theme of soil conservation recommendations for crop land, and it is also being taken up quickly in other areas, for example in southern Brazil. The application was originally in mechanized high-production farming, or for the control of wind erosion in areas of large-scale mechanized cereal production, but there are opportunities for adapting the principles and methods for small-scale farmers. The principles are equally effective in any conditions – to maximize cover by returning crop residues, not inverting the topsoil, and where rainfall conditions allow by maintaining a high crop density.

Any attempt to catalogue the methods is complicated because in some cases the same practice is referred to by different names in different countries, in other cases the same name is applied to different practices. As with most aspects of land husbandry, it cannot be too strongly emphasized that there are no practices which can be recommended for universal application and it is essential that practices must be tested under local conditions and modified where appropriate. And again we must point out that although there is sometimes a tendency to hail conservation tillage as a wonderful new development, in fact most of the methods and practices have been known and practised since man progressed from being a hunter-gatherer to a farmer who tills the soil. In his epic poem *The Georgics* completed about 30 BC, Virgil was offering excellent advice to the farmers of Italy, much of which today could be called conservation farming and including details such as different ploughing regimes for different soil types. Today there is a huge literature on the topic, from which we have selected a few references, with more detailed accounts listed in *Further reading*.

11.5.1 Surface mulching

For the conservationist any method of returning crop residues to the soil surface is beneficial because it reduces splash erosion, increases infiltration and reduces surface runoff. The agronomic or biological effects are thoroughly reviewed by RIJN (1982) and another advantage may be a reduction of soil surface temperature which can be significantly beneficial in the tropics. Other possible advantages are improvements of soil structure, fertility, and weed control.

There can also be disadvantages, particularly if the mulch is a source of weed seeds, or fosters the progressive build-up of pest or disease. Planting by machine through the mulch can be more difficult than on a clean-tilled field, but this has been largely overcome by the development of special purpose machines especially in the USA and Australia. The question of whether the mulch is better left lying on the soil surface or incorporated into the topsoil, sometimes called *trash farming*, has been extensively studied and the consensus is that the difference is usually not significant either from the point of view of yield or runoff or erosion. In Zimbabwe trials showed that incorporating the crop residues into the plough depth was equally effective when using a disc plough which results in a proportion of the crop residues being left on the surface. A similar result obtained from trials of a grass mulch applied between the rows of a coffee plantation in Kenya (PEREIRA *et al* 1964).

The spreading, chopping, and incorporation of residues is most easily achieved by machines, but the principle of mulching is equally applicable to small scale farming where the cultivation is by hand or using animals. Several long-established traditions of mulching are reported (for example HARTLEY 1938), and recent trials in many countries have demonstrated their usefulness over a wide range of farming systems and soils. Examples are in rotations of annual food crops on upland soils in Indonesia (SUWARDJO and ABUJAMIN 1985), and a range of soils growing sorghum in Ghana (BONSU 1985). A long-running research programme has operated at IITA in Nigeria (LAL 1989).

11.5.2 Reduced tillage

There are situations where conventional tillage operations such as ploughing, discing, and harrowing are suitable for preparing an appropriate seed bed, the mixing of layered soils, exposing the ploughed soil to the elements and so on almost indefinitely. However, large energy inputs are required and in today's world of farming there may be a wish to reduce these inputs, whether the objective is to reduce consumption of fossil fuels and pollution from their waste products, or whether because it is desirable to reduce expenditure of human or animal energy. The approaches to reducing the tillage requirement are many and varied and a list of the advantages and disadvantages of some of the methods popular in the mid-west of America are listed in table 11.3. The reduction can be either in time, by fewer tillage operations, or in space by cultivating only strips of land instead of the whole surface. A review of American experience comparing the yield of maize under seven alternative tillage treatments showed that most of the treatments resulted in yields not significantly different from conventional tillage but with less soil loss and runoff and with reduced inputs (DARBY 1985).

The main reduction in the energy required for ploughing came first from replacing the mouldboard plough with the disc plough, and further reduced with the introduction of the chisel plough or tine plough. For mechanized production of row crops, chisel ploughing of strips just wide enough for planting the crop (sometimes called *strip-till*) is economic and effective; an example is tobacco growing in Australia (KLEIN and PREGNO 1992).

No-till or *direct-till* is where planting is directly into the mulched soil surface, often preceded by spraying with herbicides. This method requires careful management and is mainly restricted to large-scale commercial production of crops such as cotton.

Another variation of reducing the energy input for tillage is to reduce the frequency, and this is particularly suitable to ridge tillage. The concept of building ridges and leaving them in place for several years with only minimal weeding or shaping has shown promise both in Zimbabwe (VOGEL 1992) and in the mid-western United States (BROWN *et al* 1991).

Table 11.3 Advantages, disadvantages, and typical field operations for selected tillage systems

System	*Typical field operations*	*Major advantages*	*Major disadvantages*
Moldboard plow (clean tillage)	Fall or spring plow; two spring diskings; plant; cultivate	Suited to most soil and management conditions. Fall plowing is excellent for poorly drained soils. Excellent incorporation. Well-tilled seedbed.	Little erosion control. High soil moisture loss. Timeliness considerations. Highest fuel and labor costs.
Chisel plow	Fall or spring chisel; spring disk; plant; cultivate	Less erosion than cleanly tilled systems. Less winter erosion potential than fall plow or fall disk. Fall chiseling is well adapted to poorly drained soils. Good to excellent incorporation.	Additional operations often performed result in excessive soil erosion and moisture loss. In heavy residues, stalk shredding may be necessary to avoid clogging.
Disk	Fall or spring disk; spring field cultivate; plant; cultivate	Less erosion than from cleanly tilled systems. Well adapted for lighter to medium textured, well drained soils. Good to excellent incorporation. Few residue clogging problems.	Additional operations often performed result in excessive soil erosion and moisture loss. Soil compaction associated with disking wet soils.

Table 11.3 cont. overleaf

Table 11.3 (*cont.*)

Rotary-till	Rotary-till and plant, cultivate	Excellent erosion control up to planting time. Excellent incorporation when used full width. Well-tilled seedbed.	Depending on use, low erosion control after planting; possible soil crusting; possible increased power requirements.
Ridge-plant(till-plant)	Stalk chopping; planting on ridges; cultivate to maintain ridges	Excellent erosion control if on contour. Well adapted to poorly drained soils. Excellent for furrow irrigated areas. Ridges warm up and dry out quickly. Low fuel and labor costs.	No incorporation. Creating and maintaining ridges. Keeping planter on top of ridge.
No-till (Direct drill)	Spray; plant into undisturbed surface; postemerge spraying or cultivation as necessary	Maximum erosion control. Soil moisture conservation. Minimum fuel and labor costs.	No incorporation. Increased dependence on herbicides. Not suited for poorly drained soils or weed infested fields. Management is highly critical.

(Iowa State University, Extension, AE3049, 1990)

11.6 Agroforestry

Agroforestry is not a completely new thought, but defining it as a subject in its own right, reinforced by the establishment of the International Centre for Research in Agroforestry (ICRAF), has focused attention on it and shown that it has great potential, although still at an early stage of development.

The choice of the term *agroforestry* has been questioned since it is liable to give the wrong impression that it is a branch of forestry in which some agriculture is carried out between the trees. This may have led to the unfortunate decision in some countries to place the responsibility for agroforestry within a department of forestry instead of a department of agriculture. In fact it requires a combination of agronomists and woody plant specialists, but not specialists in forests, and since the application is usually on farm land it should logically be treated as part of agriculture. The range of agroforestry technologies relevant to erosion control are summarised in table 11.4 and shown in figure 11.2, which are reproduced with permission from the excellent state-of-the-art review by YOUNG (1989). The potential of agroforestry was discussed at an international workshop in Puerto Rico in 1987 (MOLDENHAUER and HUDSON 1988). In his summary of the discussion Sanders points out that there are a number of problems to be overcome before agroforestry practices will be widely accepted by farmers. These include:

- The need to emphasize the agronomy aspects of agroforestry. For example, farmers are usually not interested in trees for timber production, such as foresters traditionally aim to produce. The farmers' requirement may be relatively thin poles for building, fencing, and for firewood.
- The need to develop agroforestry systems that farmers perceive to be both useful and within their physical and managerial capacities.
- The need to overcome problems of land tenure and land use rights in some countries. Farmers who do not have long-term tenure on their land are unlikely to be interested in planting trees.
- The need to overcome the logistics of providing suitable varieties of seedlings at a reasonable price, together with advice on planting and management of young trees.
- The need to develop new markets if the introduced agroforestry results in the production of cash crops such as fruit.
- The need for research on problems which can arise from competition for moisture or for light.

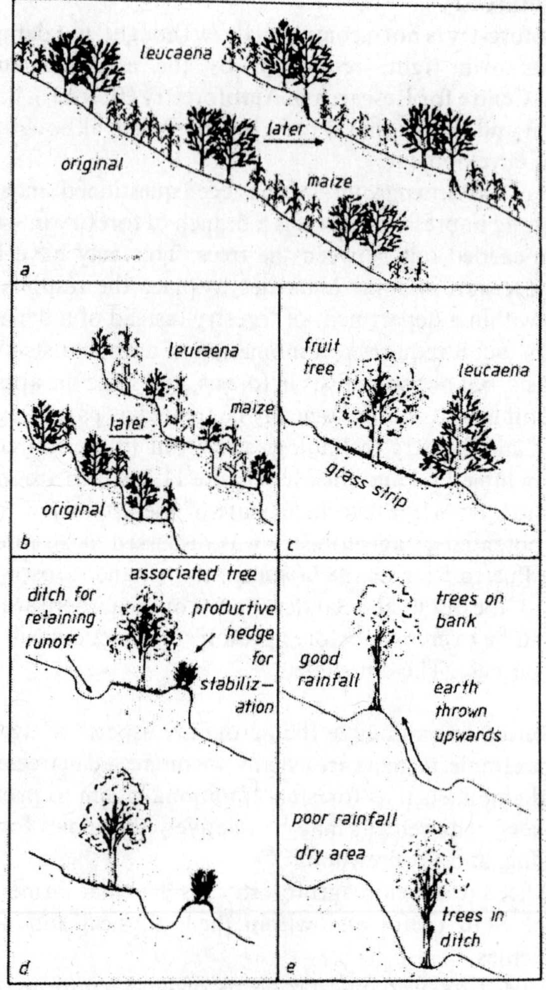

Figure 11.2 Examples of agroforestry in erosion control (YOUNG 1989)

(a) Hedges of double rows of *Leucaena* with maize between, developing into terraces, Philippines
(b) *Leucaena* hedges planted between rows of maize developing into terracettes, Malawi
(c) Trees on conservation works: fruit trees on grass strips and *Leucaena* on marker ridges to guide cultivation ridges below, Malawi
(d) Two arrangements for trees on conservation structures, Cameroon
(e) Alternative positions for trees on fanya juu structures, Kenya

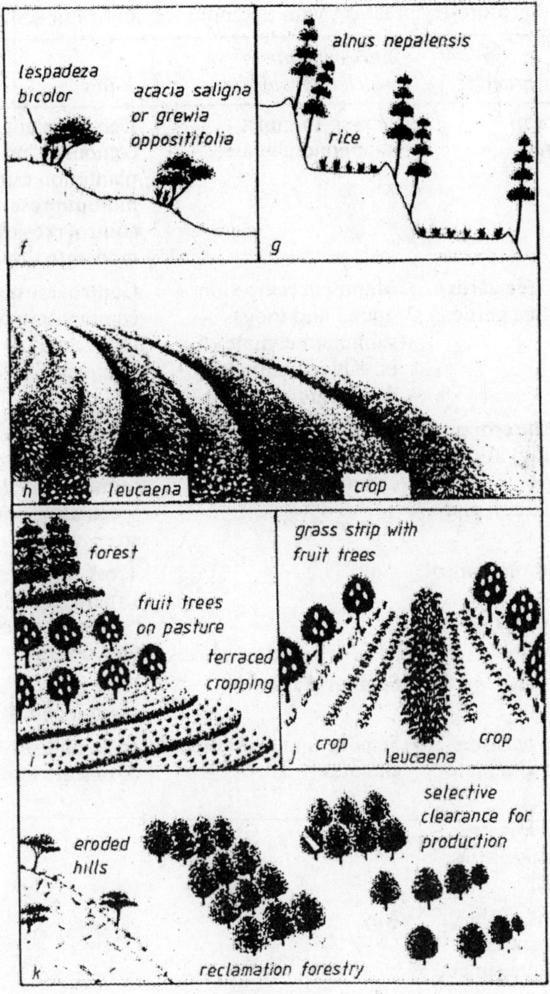

(f) Trees on terrace risers, Ethiopia
(g) Trees on risers of irrigated terraces, Nepal
(h) Hedgerow intercropping with *Leucaena* laid out on a slope, Philippines
(i) Model for land use as an alternative to shifting cultivation, India
(j) Combining barrier hedges with trees on grass barrier strips, Philippines
(k) Possible development of reclamation forestry into productive use by selective clearance of contour strips

Table 11.4. Agroforestry practices with potential for control of soil erosion

Agroforestry practice	*Environments in which applicable*	*Notes*
Plantation crop combinations	Humid to moist subhumid climates	Densely planted combinations of plantation crops with multipurpose trees control erosion on moderate slopes
Multistorey tree gardens, including home gardens	Mainly developed in humid and moist subhumid climates but possible potential in drier regions	Control erosion through combination of herbaceous cover with abundant litter
Hedgerow intercropping (alley cropping) and barrier hedges (figure 11.2 a, b, h and plate 11.3)	Humid, subhumid and possibly semi-arid climates	Combine erosion control with arable use on gentle to moderate slopes; more speculative potential on steep slopes
Trees on erosion-control structures (figure 11.2 d, e, f, g)	Any	Trees stabilize earth structures and give production from land they occupy
Windbreaks and shelterbelts	Semi-arid zone	Well known as wind erosion control measures
Silvipastoral practices (figure 11.2 i and plate 11.4)	Semi-arid and subhumid climates	Trees and shrubs as part of pasture improvement
Reclamation forestry leading to multiple use (figure 11.2 k)	Any	
Combinations of the above in integrated watershed management	Any	

From *Agroforestry for Soil Conservation* (YOUNG 1989)

A wide range of species of trees and shrubs have been used in agroforestry and there are several compilations of possible species, sometimes specific to a particular region such as ROCHELEAU *et al* for Africa (1988). A review of the potential for agroforestry, and of suggestions for both research on the techniques and on treating agroforestry in a multi-disciplinary and inter-institutional way, were reviewed by LUNDGREN and YOUNG (1992). Two examples which are proving particularly popular are the use of *contour hedges*, either for erosion control as in Haiti (PELLEK 1992) or the double hedgerow method developed in the Philippines as part of the Sloping Agricultural Land Technology (SALT) reviewed by TACIO (1991). The other popular application is *alley cropping* or growing foodcrops between hedges which can be used to gain limited food production on land which is too steep for other forms of arable land use, as practised in the highlands of northern Thailand (PREECHAPUNYA 1987) (plates 11.3 and 11.4). Continuing research at IITA in Nigeria is demonstrating the long-term advantages of alley cropping using hedgerows of *Leucaena* and *Gliricidia* (KANG and GHUMAN 1991).

References

BONSU, M. 1985 Organic residues for less erosion and more grain in Ghana, chapter 58, in *Soil Erosion and Conservation*, edited by S. A. El-Swaify, W. C. Moldenhauer and A. Lo, Soil Conservation Society of America, Ankeny, Iowa

BROWN, L. C., L. D. NORTON and R. C. REEDER 1991 Erosion and Yield Benefits from Ridge-Tillage Systems, in *Proceedings of 12th International Conference of International Soil Tillage Research Association*

DARBY, G. M. 1985 Conservation Tillage: an important adaptable tool for soil and water conservation, chapter 62, in *Soil Erosion and Conservation*, edited by S. A. El-Swaify, W. C. Moldenhauer and A. Lo, Soil Conservation Society of America, Ankeny, Iowa

FAO 1987 Soil and Water Conservation in Semi-Arid Areas, *Soils Bulletin 57*, FAO, Rome

HARTLEY, B. J. 1938 An indigenous system of soil protection, *East African Agriculture Journal*, 4, 63–66

HUDSON, N. W. 1957a Soil Erosion and Tobacco Growing, *Rhodesia Agriculture Journal*, 54, 6, 547–555

HUDSON, N. W. 1957b Erosion Control Research, *Rhodesia Agriculture Journal*, 54, 4, 297–323

IOWA STATE UNIVERSITY 1990 *Conservation Tillage*, Iowa Cooperative

Plate 11.3 Double rows of *leucaena* for mulch or fodder. An annual arable crop will be planted between the rows in the rainy season (ICRISAT)

Plate 11.4 A silvipastoral experiment in India. Grass between rows of mature *leucaena* is cut for stall feeding of cattle. The soil is shallow gravel and would not support a food crop (ICRISAT)

Extension Service, Iowa State University, Ames, Iowa, Publications AE3049 to AE3057

KANG, B. T. and B. S. GHUMAN 1991 Alley cropping as a sustainable system, in *Development of Conservation Farming on Hillslopes*, edited by W. C. Moldenhauer, N. W. Hudson, T. Sheng and San-wei Lee, Soil Conservation Society of America, Ankeny, Iowa

KLEIN, J. F. and L. M. PREGNO 1992 Reduced Tillage in Tobacco, in *People Protecting their Land*, edited by P. G. Hoskins and B. M. Murphy, Department of Conservation and Land Management, Sydney, New South Wales

LAL, R. 1989 Conservation Tillage for Sustainable Agriculture: Tropics versus Temperate Environments, in *Advances in Agronomy*, 42, edited by N. C. Brady, 85–197, Academic Press, New York

LUNDGREN, B. and A. YOUNG 1992 Land Use Management in relation to soil conservation and agroforestry, in *Erosion, Conservation, and Small-scale Farming*, edited by H. Hurni and K. Tato, University of Berne, Switzerland

MANNERING, J. V. and C. R. FENSTER 1983 What is conservation tillage? *Journal of Soil and Water Conservation*, 38, 3, 141–143

MOLDENHAUER, W. C. and N. W. HUDSON (editors) 1988 Conservation Farming on Steep Lands, *Soil and Water Conservation Society*, World Association of Soil and Water Conservation, Ankeny, Iowa

PELLEK, R. 1992 Contour hedgerows and other soil conservation interventions for the hilly terrain of Haiti, in *Erosion, Conservation, and Small-scale Farming*, Vol. 2, 29, 313–320, edited by K. Tato and H. Hurni, University of Berne, Switzerland

PEREIRA, H. C., M. DAGG and P. H. HOSEGOOD 1964 A tillage study in Kenya coffee, Part IV, The physical effects of contrasting tillage treatments over thirty consecutive cultivation seasons, *Empire Journal of Experimental Agriculture*, 32, 125

PREECHAPUNYA, P. 1987 Agroforestry systems for highland watershed development in Northern Thailand, in *Steepland Agriculture in the Humid Tropics*, 675–703, Malaysian Agriculture Research and Development Institute (MARDI), Serdang, Malaysia

RIJN, P. J. VAN 1982 No-tillage crop production in the tropics. *Abstracts in Tropical Agriculture*, 8, 3, 9–27

ROCHELEAU, D., F. WEBER and A. FIELD-JUMA 1988 Agroforestry in Dryland Africa, *Science and Practice of Agroforestry 3*, ICRAF, Nairobi, Kenya

SHAXSON, T. E., N. W. HUDSON, D. W. SANDERS, E. ROOSE and W. C. MOLDENHAUER 1989 *Land Husbandry: a Framework for Soil and*

Water Conservation, Soil and Water Conservation Society, Ankeny, Iowa

STEWART, J. I. 1988 *Response Farming in Rainfed Agriculture*, Wharf Press, Davis, California

STOCKING, M. 1993 Soil and water conservation for resource-poor farmers: Designing acceptable technologies for rainfed conditions in eastern India, in *Topics in Applied Resource Management in the Tropics, Vol. 3, Acceptance of soil and water conservation: Strategies and Technologies*, German Institute for Tropical and Subtropical Agriculture (DITSL), Witzenhausen

SUWARDJO and S. ABUJAMIN 1985 Crop residue mulch for conserving soil in uplands of Indonesia, chapter 57, in *Soil Erosion and Conservation*, edited by S. A. El-Swaify, W. C. Moldenhauer and A. Lo, Soil Conservation Society of America, Ankeny, Iowa

SWAMINATHAN, M. S. 1982 *Rainfall and Dryland Agriculture*, Proceedings of a symposium on Rainwater and Dryland Agriculture, Indian Agricultural Research Council, New Delhi, October 1980

TACIO, H. D. 1991 The SALT System: agroforestry for sloping lands, *Agroforestry Today*, 3, 1, 12–13

VOGEL, H. 1992 Conservation Tillage for sustainable crop production – concept and approach of a development programme in Zimbabwe, in *People Protecting their Land*, edited by P. G. Hoskins and B. M. Murphy, Department of Conservation and Land Management, Sydney, New South Wales

WILLCOCKS, T. J. 1984 Tillage requirements in relation to soil type in semi-arid rainfed agriculture, *Journal of Agricultural Engineering Research*, 30, 327–336

YOUNG, A. 1989 Agroforestry for Soil Conservation, *CAB International*, Wallingford, Oxfordshire

12

Gully erosion

12.1 The nature of gully erosion

Gully erosion is spectacular and widespread, and so it is often used by conservationists as a characteristic symptom of erosion. As a result there is a danger of its importance being over-emphasized. Certainly it is most important as a source of sediment in streams, but in terms of damage to agricultural land or reduction of agricultural production, it is usually not very important for the simple reason that most land subject to severe gully erosion is of little agricultural significance. The really spectacular erosion so popular in non-technical reports shows large areas cut up by deep interconnecting gullies, but this situation is usually found in semi-arid climates unsuitable for any serious agriculture, or on soils which have such adverse chemical or physical properties that their potential production is very low. Added to this is the fact that gully control is always difficult and expensive, so the cost of reclamation usually exceeds the value of the land. Certainly in the sense of reducing the amount of eroded soil choking the dams and rivers it may be highly desirable to do something about gully erosion, but it is very much a case of prevention being better than cure. Limited resources in manpower, money and materials can usually be better employed in preventing future gullies than in curing existing ones.

The worldwide occurrence and recognition is well demonstrated in the variety of names: in English and American – *gully*, in Egypt – *wadi*, in South Africa – *donga*, in French-speaking lands – *ravine*, in India – *nulla*, in South America – *carcava*, or *arroyo*, and many others. The definition is the same in any case, a steep-sided eroding watercourse which is subject to intermittent flash floods. They are more common in semi-arid climates with infertile soils and sparse vegetation, but some spectacular examples are found in tropical forests in deep soils with dense vegetation.

Plate 12.1 Gullies in New South Wales, Australia, caused by increased runoff from overgrazing by rabbits, stabilized naturally when the rabbits were destroyed

Attempts have been made to identify the factors which affect the growth of gullies (STOCKING 1980) and also to predict situations when gullying is likely to occur (IMESON 1980). However, the wide variations of conditions make it difficult to turn our qualitative understanding of the process of gullying into quantitative terms. Quantifying the soil loss from gullies is also far from precise, in spite of experimental studies such as that of PIEST *et al* (1975). Gully control must start by identifying the causes, as demonstrated by an unusual example from New South Wales, Australia. An exploding population of rabbits resulted in overgrazing and denudation of grassland causing an increase of runoff which started gullying in previously stable streams. When the rabbits were wiped out by an epidemic of the contagious disease myxomatosis, the flooding stopped and the gullies stabilized (plate 12.1).

12.2 The causes of gully erosion

The common factor in all cases of gully erosion is that the basic cause is the same, and the whole cycle of gully formation and growth is clear and understandable once the cause is defined. In engineering terms the cause is the breakdown of a state of metastable equilibrium in the stream or watercourse. In physics a body is said to be in equilibrium if it will return to the same position after being displaced slightly, for example if a pendulum is pushed to one side it will return under the force of gravity to its position of equilibrium. *Metastable equilibrium* is when the body will only come back to its equilibrium position if it not displaced too much. A match box standing on one end can be tilted slightly and it will return to its original position, but if it is displaced too far it falls over and the equilibrium is lost. The second feature of metastable equilibrium is that in order to restore the original position it is necessary to apply a force which is greater than the force which toppled the body over in the first place.

Gully erosion occurs when a natural watercourse is displaced from its state of metastable equilibrium. A watercourse is ordinarily in balance, that is the size of the channel and its shape and gradient are on the whole suitable for the flows it has to carry. If the balance is slightly changed by an external force the stream will tend to find a new equilibrium. For example, if the flow increases the channel may enlarge itself, or the gradient might increase, until a new balance is found. But the stream, like the match box standing on one end, can only come back to its original position after small disturbances. If given a hard push the matchbox falls over, and if given too big a change the stream starts gullying, and once the starting position has been lost, then it requires a greater effort to get back to it.

The disturbing influence which upsets the balance of a watercourse may result from either of two sources. Either there is an increase in the amount of flood runoff which the channel has to carry, or the flood water remains the same but there is a decrease in the ability of the channel to carry that flood.

12.2.1 Examples of increased floods

The most common causes of an increase in flood runoff are changes in land use. In a heavily timbered catchment, if there is much cutting of the trees, then the maximum flood flow will usually increase. Increases in the proportion of arable land in a catchment could have the same effect, and so could excessive burning of the vegetation or overgrazing (plate 12.1). All of these have caused innumerable gullies by increasing the flood runoff so much that the watercourse could not adjust and lost its metastable equilibrium.

The maximum flood flow will be increased if the catchment area is increased, and two examples of this can be illustrated. When major roads are constructed this often changes the natural drainage patterns and the natural catchments. Figure 12.1 shows a commonly occurring situation. Several small catchments each have their own watercourse carrying flood

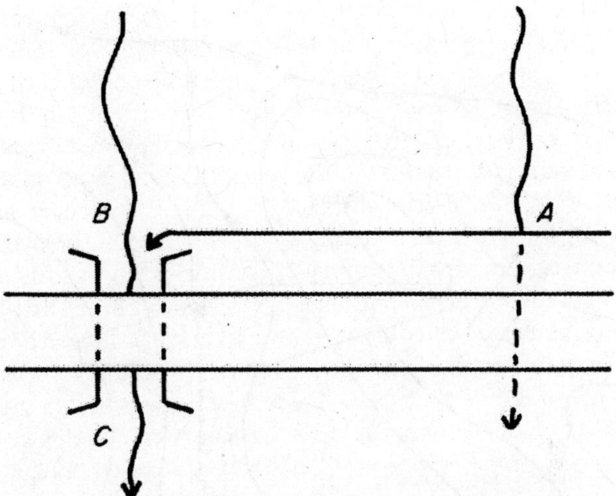

Figure 12.1 Gully erosion can be started by changes in road drains

flows, but when the road is built it would be expensive to carry each minor stream under the road by means of a bridge or culvert. A common method is to close one watercourse (*A* in figure 12.1) where it crosses the line of the road, and to divert the runoff along the road drain and add it to that of the next watercourse (*B* in figure 12.1), passing the combined flow through a culvert. Where the culvert discharges, the watercourse is capable of handling the flow from catchment *B*, but is quite unable to cope also with the added flow from *A*, and gully erosion starts at point *C*.

Another example of artificially increasing the catchment area is when flood runoff is by-passed into an adjoining catchment as the result of the construction of a dam. In figure 12.2, catchment *A* is made into a reservoir and a convenient way of passing the flood flows when the dam is full is to spill the water over the ridge on the left bank and into the adjoining catchment. The stretch of stream from point *B*, where the flood flows now enter this stream, down to point *C*, the original confluence, is now being asked to carry a greater flow than previously. If the increase in the catchment is large there is a strong probability that the equilibrium will be upset leading to gully erosion.

12.2.2 Examples of decreased channel capacity

In chapter 8 the capacity of a channel was shown to depend on the physical characteristics of the channel, like cross-sectional area, shape,

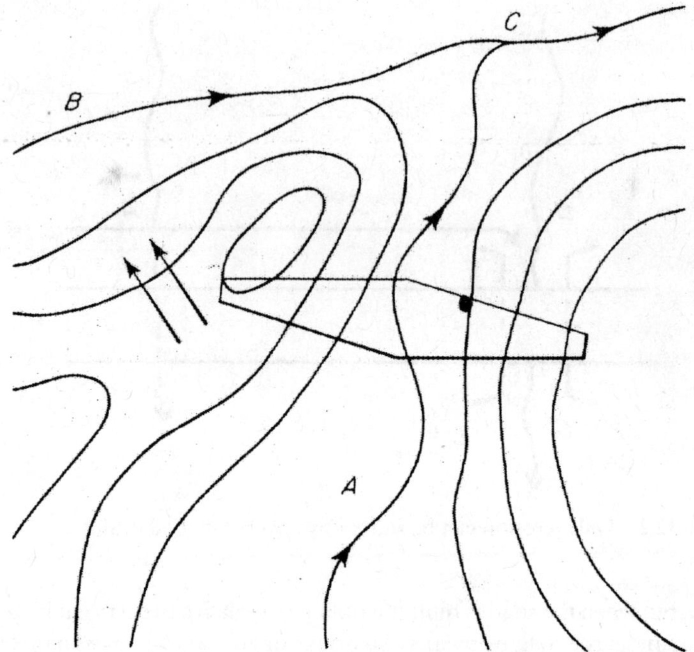

Figure 12.2 An example of a change in the catchment area

gradient and roughness. Changes in these channel factors can easily upset the equilibrium. For example, the nature of the vegetation in the channel has a big influence on the velocity of flow. An increase in the density of vegetation could result from natural or man-made causes. Perhaps the stream banks have been planted with trees or shrubs and these are now growing well, or perhaps the natural grass has been replaced with a more vigorous type, or the grass is increasing as a result of fencing which now keeps out livestock. Any of a dozen such reasons could result in a denser or taller vegetation, with more resistence (i.e. higher values of roughness coefficient n). If flood waters cannot flow away quickly enough down the channel, they may spill over the banks and start new erosion patterns.

On the other hand a reduction in the density of vegetation may also be the trigger which sets off gullying. Overgrazing or fire are probable causes, and the effect is a lowering of the hydraulic friction, and an increase in velocity, with the risk of scour erosion starting.

The capacity of a channel is also affected by its shape, which can be defined numerically by the term *hydraulic radius* (section 8.1.1). Any changes in the shape may affect the capacity, and again these may be natural or man-made or a combination of both. The deposition of sand bars or the collapse of streambanks will result in lower values of hydraulic radius, and lower velocity, and hence reduce channel capacity. On the other hand scouring of the bed will give increased capacity, but the velocity will also increase and so the scouring action may continue. Small localized problem areas in a watercourse can have a significance out of proportion to their size by starting a cycle of events which expand and multiply. There are countless examples of large gullies which have resulted from insignificant beginnings like a small cattle track.

The concentration of flood waters into restricted channels is another potential cause of gullies. An example is a road which crosses a natural stream. Before the road is built, any flood waters can spread out over the whole of the valley, in fact if the stream is prone to floods there will probably be a flood plain which is part of the balanced equilibrium which the stream has achieved over the years. When the road is built there will be a bridge, or causeway, or culvert, to pass the floods but whichever of these is used there will be some restriction of the cross-sectional area available. The flow will be concentrated into a smaller area so the velocity will increase, and there is a good chance of gully erosion developing immediately downstream of the crossing. Large streams and rivers are likely to have cut a fairly stable channel down to rock or hard earth, and so the danger of this type of erosion is less serious in the case of main roads or railways crossing major rivers. But when a lightly trafficked farm road is put through a small watercourse which carries only intermittent floods, then the danger is much greater.

Another case of concentration which can lead to scouring is the design of earth spillways from small conservation dams. A common practice is to pass the overflow back to the stream by a cut channel round the end of the dam wall. This part is usually properly designed to cater for the maximum estimated flood, but the water then leaves the cut spillway and runs down the bank to the stream bed. Here there is a concentration of flow over ground quite unused to it, and a gully which starts at this point can quickly cut back until it empties the dam and makes it useless.

The progressive action of gullies, where once the balance is upset the problem gets increasingly worse, is also seen in the longitudinal profile of the floor of gullied streams. For a stream in balance (sometimes referred to as 'in regime') the stream gradient and the roughness are two of the interrelated factors. *Manning's formula* links them with velocity in the form:

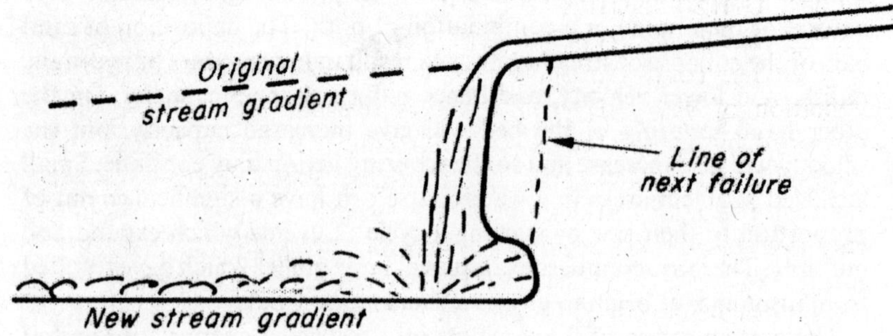

Figure 12.3 The progressive development of the head of a gully

$$V = \frac{R^{\frac{2}{3}} S^{\frac{1}{2}}}{n}$$

where V is the average velocity of flow in metres per second
R is the hydraulic radius in metres (the letter M is also used to denote hydraulic radius)
S is the average gradient of the channel in metres per metre (the letter i is also used to denote gradient)
n is a coefficient, known as *Manning's n*, or *Manning's roughness coefficient*

Once gullying starts the gullied channel has a more angular and deep shape than the original bed, i.e. R increases. The gullied channel is bare of vegetation so the roughness coefficient, n, probably decreases. For the velocity to remain constant the gradient must therefore decrease, and this is what almost invariably happens; the gradient of the floor of the gully is flatter than the original stream. The result, as shown in figure 12.3, is that as the head of the gully works back upstream the height of the overfall increases. This overfall is usually the most actively eroding part of the gully. The waterfall action both scours the soil where it lands and also it splashes and swirls against the face of the waterfall. The bottom of the face is eroded away leaving the top overhanging. The overhang breaks away to give a vertical face, then the cycle starts again.

Reverting to Manning's formula, the changes, once gullying starts, are that hydraulic radius, R, will increase, roughness coefficient, n, will decrease, and the gradient, S, will probably decrease. The overall effect is most likely to be an increase in velocity, and this is yet another reason why gully erosion is more often self-perpetuating than self-correcting.

12.3 Control measures

12.3.1 The economics of gully reclamation

The control of gully erosion is difficult and expensive. The definition of *metastable equilibrium* is relevant: once the balance is disturbed by an upsetting force something is irretrievably lost and it requires a greater force to restore the original state. 'Prevention is better than cure' is a trite saying but very appropriate to gully erosion. The restoring force usually requires a great deal of time, effort, and money, so much so that often the same effort would be better employed in preventing new gullies from starting, rather than trying to cure or restore existing ones. In terms of simple economics, the repair of gullies is seldom justified since the cost of repair is likely to be higher than the value of the land after it has been reclaimed, particularly as gullies are most common in semi-arid conditions with poor soils of low agricultural value. However, other factors have to be considered before deciding whether control work is justified. Quite apart from the on-site damage, it may be necessary to do something about the erosion because of downstream effects such as a storage dam being silted up or irrigation works threatened. Or there may be on-site damage other than the obvious loss of land. Examples are the lowering of the water table, the interface with fences or roads, or perhaps the possibility that if allowed to continue unchecked the gully may in time work back to a road, a bridge, or a building. All these factors must be considered but they can seldom be evaluated in terms of actual cash, and cost-benefit analysis techniques do not help. In fact they are difficult to apply to any aspect of conservation work because there is no scale of values with which to measure the cash value to the community of benefits, such as not lowering the water table (section 14.4). In practice what happens is that the economic overlords rule that schemes may only be undertaken if the ratio of benefits to costs exceeds a certain value, say one and a half. The conservationist then decides on the basis of his knowledge and experience which works should be done, and calculates the cost. He then multiplies the cost by the required ratio and uses this figure to make up his estimate of the benefits. When the scheme is submitted it meets the cost benefit requirement and is approved. In this way the right works are built, the administrators are kept happy, and the correct use of public funds can be demonstrated, but one cannot help feeling that there ought to be a better way.

One form of gully control which may be economically sound but hard to achieve is partial control, that is where the object is not to reclaim or restore the gully, but merely to prevent it getting any worse. This may be required because of on-site problems, for example the gully head is working back each year, and threatens a road bridge. Alternatively it may be desirable to prevent further sediment load in a stream, but the land already gullied is not worth reclaiming. A possible solution would be a cheap simple structure at the gully head and nothing more, but unfortunately partial control is like trying to find an intermediate position of equilibrium half way back to the original state of metastable equilibrium. In the example of the match box, when it is standing on end it is in metastable equilibrium. The collapsed state, corresponding to the gully, is when the box is lying flat. What is not possible is to balance it on one corner in a half-way position. It is almost as difficult to achieve a half-way balance in gully control, and partial control measures seldom work except temporarily.

12.3.2 Principles of gully control

The reason why partial control is unsuccessful is that it does not conform to the first principle of gully control which is to determine the cause of the gully and to take counter measures. A doctor does not start trying to cure an illness until he knows the nature of the disease and its cause. If the stability has been lost and the gully started because the volume of flood water has doubled then minor patching up of the damage will not solve the problem. The second principle is either to restore the original hydraulic balance or to create new conditions. Either the flood has to be reduced to its original volume or a new channel has to be provided which can accommodate the increased flood. Some interesting experiments on vegetated channels for this purpose have been carried out in America (HEEDE 1987).

In most cases it will be found desirable to fence off any area where control measures are being applied. Whether the controls are structures or plantings they are vulnerable while being established, and interference by grazing animals, or children playing, or from any other source must be minimized.

12.4 Control by vegetation

12.4.1 Why use vegetation?

A conservation engineer of many years experience is reported as having once said: 'In gully control a bag of fertilizer is more effective than a bag of cement'. He would no doubt have amplified this by explaining that although structural works are sometimes necessary, it is on the whole preferable to restore gullies by the use of vegetation. Structures, whether of concrete, masonry, wood or any other building material, are subject to decay, and liable to be undermined or bypassed. They can only become less effective with the passing of time. Vegetation, on the other hand, can multiply and thrive and improve over the years. Structures also need various skills for their design and construction, and they are usually expensive.

The purpose of vegetation is twofold: it provides the soil with physical protection against scour, and it slows down the velocity of flow by increasing the hydraulic resistance of the channel and thereby greatly reduces the scouring and abrading ability of the flood. If the velocity is sufficiently reduced then some of the sediment load may be deposited. This can lead to the desirable situation where the vegetation, growing vigorously in the deposited soil, gets denser and so traps more and more silt until the whole gully is filled. Of course this can only happen where the source of the sediment is erosion higher up the catchment.

However, there are considerable obstacles to the establishment of vegetation in gullies. The environment is usually just about as inhospitable as it could be. The bed of the gully is probably almost sterile sand with no structure, no organic matter, no available plant nutrients, and low moisture-holding capacity. If there is any moisture it is most likely to be deep down, well below the rooting range of any newly established plants. Yet this is precisely where grasses, reeds, and sedges would be most useful. The banks of streams and sides of the gully are probably not much better. The water regime is alternations between flood and drought, and problems of chemical imbalance such as salinity and alkalinity are common in soils subject to gullying (HUDSON 1963).

12.4.2 Plants and planting techniques

The two ways of overcoming these problems are the selection of suitable plants and the use of special planting techniques. Using plants which have proved successful in other countries is always worth trying but may be disappointing. A good example of this is Kudzu vine, (*Pueraria thunbergiana*), which has been successfully used in many parts of the United States. Given the right conditions it puts out vigorous creeping runners which can completely cover the floor, sides and flanks of a gully with a dense blanket of vegetation. However, its performance has been most disappointing in several countries in Africa, though the explanation for this is not known. It is seldom possible to predict which plants will do well, and a useful practice is to make small trial plantings of a wide range of plants to see which flourish. Appendix 2 lists some plants which have been found useful in various parts of the world, and a useful review of plants used in America is given by BENNETT (1939). In the search for suitable varieties, local material should not be overlooked. Any plants which are growing reasonably well in or near the gully must be accustomed to local conditions. Sometimes a little help in the way of fertilizer will enable them to outgrow all the exotic varieties imported from other countries. The requirements for plants to be suitable for gully control are that they should grow vigorously in poor conditions and give good ground cover. A spreading, creeping habit is much better than an upright habit.

Plants are usually best established by planting out seedlings which have been germinated and started in a more favourable medium. One popular method is to use cylinders in thin black polythene to form bottomless plant pots. These are filled with good soil and the seedling makes good growth. When planting out, a cylindrical hole is made with a soil auger and the plant and soil placed in the hole. The plastic film can be left on to hold the soil in place. By the time the plant has outgrown its reservoir of fertile soil it is strong enough to survive in the tougher conditions outside. This method might be used to establish a clump of *Phragmytes* reed, planting out at half-metre intervals. A variation on this idea, useful for establishing colonies of grass, is to plant into sacks of good soil (DURBACH 1964). The sacks are laid in shallow trenches in the bed of the gully so they are about level with the bed. A small cut is made in the bag and a seedling planted through the cut. The bag prevents the soil and plant being swept away by the first flood, and by the time the bag rots away the plants have made sufficient root growth and no longer need the protection of the bag. Old jute or hessian sacks are best because the grass can grow through them, but discarded plastic fertilizer bags, or strong paper sacks, can also be used.

Planting the sides of gullies is difficult, because they are steep, unstable, and eroding. When the cost is justified the banks can be levelled to a gentle uniform slope by heavy earthmoving machinery, and then seeded or planted. Sometimes sufficient soil is bulldozed in from the sides to convert the gully into a shallow grassed waterway (HARRIS and HAY 1963). Since the banks are almost sure to be infertile subsoil, some extra fertility must be added. One method is to insert pockets of better soil as was described for planting the floor of the gully. Alternatively it is sometimes possible to bulldoze a pile of top soil to the edge of the gully and then spread it as a blanket over the sloping sides of the gully. This can then be seeded or planted, but there is the problem that soil or seed or planting material all tend to get washed off the slopes before the vegetation can become established. Surface mulches of straw or other crop residues, or long grass cut for the purpose, are valuable in accelerating the establishment of vegetation, but the mulches too are prone to surface wash. In the United

States of America several man-made materials are produced especially for anchoring mulches and planting material. One form is a coarse open-mesh woven net and this is made from jute fibre and also from kraft paper. Another form is made from fine wood shavings and known under the trade name *Excelsior*. Examples are shown in plates 13.2 and 13.3. The netting is unrolled over the top of the mulch and then fastened down with wire spikes pushed into the soil. The method is undoubtedly very effective and although the cost of importing these materials from America is prohibitive, there seems to be no reason why cheap substitutes could not be manufactured locally, especially in countries producing large quantities of cloth fibres. Some other methods of stabilizing sloping banks are discussed in section 13.2.3 in connection with road embankments.

12.5 Structures for gully control

12.5.1 Temporary structures

It frequently happens that the establishment of vegetation is difficult because the newly planted material gets swept away, or because there is no soil for the vegetation to grow in. In either of these cases there may be a place for temporary structures whose purpose is to provide protection for just long enough to give vegetation a start. If the object is to slow down the water and so cause deposition of silt there is no need for the structures to be watertight, and the term *porous checks* describes this type.

Wire bolsters

A simple but effective method, if there is plenty of loose rock available nearby, is to build a loose rock-fill dam with the stones anchored in place by wire netting. Galvanized wire netting of a fairly stout gauge and two metres or more in width is laid out flat across the gully bed. Loose rock is packed on one half of the width of the netting and the other half is wrapped over the stones and laced to the other edge, forming a sausage or bolster of rock contained in a skin of wire netting. More substantial structures can be built using several layers of rock bolsters (HEEDE 1977). The technique was recently used successfuly in Mali by RANDS (1992). A more sophisticated commercial product is described under permanent structures in section 12.5.2.

Netting dams

Another use of wire netting is to form small check dams, usually near the top end of gullies. Wooden posts are driven into the bed of the gully and used to support a strip of wire netting which forms a low wall across the gully. The height should be only a half metre or so and the lower edge of the netting is buried. Light brush or straw is piled loosely against the upstream side of the netting wall and is packed by the flow of water against the netting to form a barrier, which is porous but slows down the flow and causes a build-up of sediment on the upstream side.

Brushwood dams

The main requirement of temporary control structures is that they must be quick and easy to construct, and use cheap readily available materials. In wooded areas two types of silt retaining dam are used. The brushwood dam, shown in figure 12.4, uses small branches, up to 2 or 3 cm in diameter, packed as tightly as possible across the direction of flow. They can be anchored by packing them between rows of vertical stakes, or by tying down with wire, or by sticks laid across the top and fastened down. The main points in building these dams are to pack the brushwood as tightly as possible and to secure it firmly. With attention given to both these points it is not uncommon for brushwood check dams to last for several years.

Log dams

When heavier timber is available it can be used for log-piling dams (HEEDE 1977). One method is to use logs in the same way as the brushwood dam but to make a much more substantial structure. Two rows of vertical posts are driven into the bed of the gully and extending up the sides to above flood level, and then logs are packed in between. The vertical posts should be at least 10 cm diameter, 2 m long, and spaced about a metre apart in each row, with the two rows of posts half a metre apart. In a wide shallow stream it is best to drive in all the vertical posts to the same height above ground, about half a metre, so that the top of the

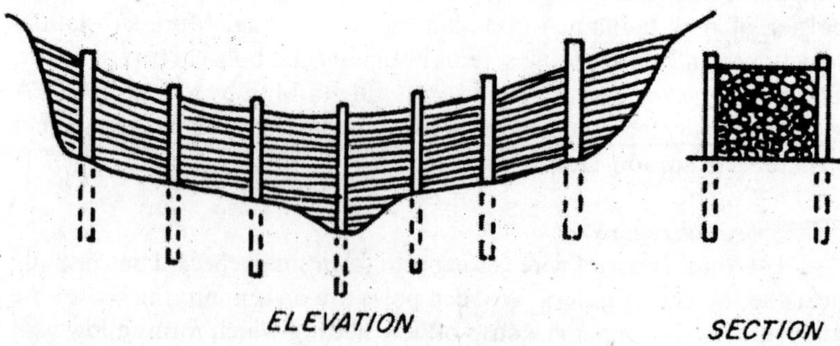

Figure 12.4 A brushwood dam for gully control

dam follows the section of the stream bed (figure 12.4). If the gully has steep sides it is better to have a rectangular notch in the centre (figure 12.5) but the notch must be big enough for the whole of the flood to pass through. A common error in the construction of this kind of dam is to make the notch too small so that the floods go over the top. This is very likely to start scouring at the sides of the wall. The vertical posts on either side of the notch will carry the brunt of the force of the flood waters and any logs or boulders swept down by the flood, so they should be particularly stout posts and driven in deeper than the others. When the logs are packed between the rows of posts the bottom layer should be sunk below ground surface to avoid seepage and scour underneath. After the top logs have been placed they are held in position by strong wire ties between the vertical posts.

A simpler structure can be made, consisting only of a single row of vertical posts driven in side by side to form a wall of logs. Again they can follow the profile of the gully section or have a central notch. This structure depends on the firmness with which the posts are held in the bed, so is most suitable where long posts can be driven deep into a firm soil. Some extra rigidity is obtained by lashing or bolting a few cross-members to the vertical posts.

Brick weirs

The dividing line between temporary and permanent structures is arbitrary; many structures could have a very variable life depending on how they were constructed and maintained, and what pressure of use they had to withstand. Sometimes long-lasting materials are used in structures with a short design life, for example brick weirs designed to hold sediment

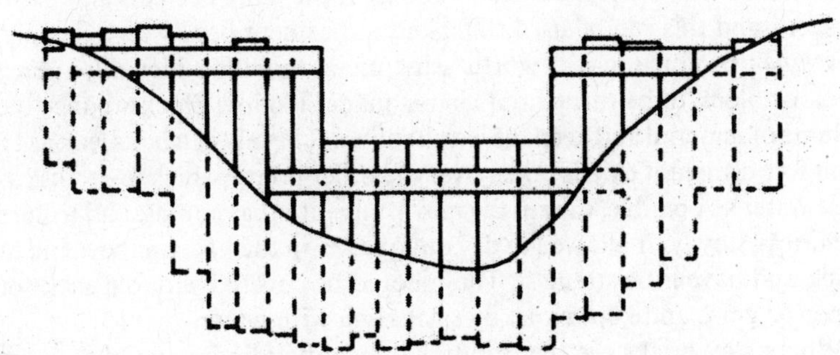

Figure 12.5 Timber piling used to make a log dam for gully control

Plate 12.2 A simple brick weir, built up in several stages as the silt accumulated on the upstream side. The notch is much too small

or water long enough to establish vegetation. Some examples are shown in plates 12.2 and 12.3 from HUDSON (1963).

Since the object is cheapness and simplicity the materials and the design must be chosen to suit the site conditions. If the gully bed contains clean washed sand this can be used to make sand/cement bricks very cheaply. For quick production it is worth using metal moulds which allow each brick or block to be turned out as it is made. Hollow blocks reduce the volume of material and keep the weight down. The mix can be as lean as 15 sand to 1 cement if care is taken over the other points which are to use as little water as possible, to ram the mix firmly into the moulds, and to cure the bricks slowly. If allowed to dry out quickly in the hot sun they tend to crack and have little strength. They should be covered with old sacks or paper or grass and kept moist by sprinkling with water.

Where clay bricks are the traditional local building material they are probably cheap enough for use in gully control structures. Bricks burned

Plate 12.3 A simple arch weir built from sand-cement bricks. There are no foundations and the stresses are carried by the rock outcrop on each flank

in a kiln are normally resistant to the effect of water but sun-dried bricks will stand occasional wetting and may be used if the gully only experiences infrequent floods. Second-grade bricks or rejects not good enough for building but quite adequate for gully control can often be obtained very cheaply from commercial brickfields.

Some simple designs are shown in figure 12.6. The shape which gives the best strength/weight ratio is the arch weir, illustrated in plate 12.3, and a single thickness of brickwork can be built to a height of one or one and half metres over a circular span of about two metres. A straight wall of similar size would need three or four times as much brickwork to achieve comparable strength. The arch wall works by transmitting the load round the arch to the buttresses at each end, and so it needs good solid support in the gully ways, preferably in the form of a rock outcrop. However, the light wall does not need much in the way of foundations, and in fact the weir shown in plate 12.3 is resting only on a bed of sand between the rocks on either side.

In the more usual situation of a rock bar, which runs across the bed of the gully, a straight gravity section wall is indicated. The width at the base should be approximately equal to the maximum height, and successive courses of brickwork are narrower so that the section is roughly triangular. It is common to find the upstream face of dams vertical, with

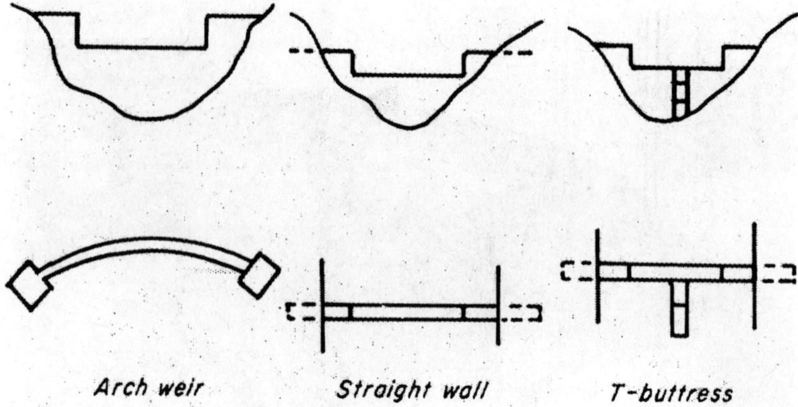

Figure 12.6 Types of small brick weirs for gully control

all the slope on the downstream face, but while there is a sound engineering reason for this in the case of large water storage dams, it is not of any consequence in small gully control dams.

Since brickwork has little tensile strength the weakest feature of straight-wall brick dams is their resistance to the bending moment which results from the water pressure. The bending moment M is related to the length of wall l, and the load per unit length w, by the formula $M = wl^2/8$. In other words the risk of failure is proportional not to the length of the wall, but the square of the length. A buttress at the mid-point of a straight wall reduces the effective length by a half but cuts the bending moment to a quarter so is usually worthwhile whenever the site conditions allow. Some other construction points are discussed in the following sections on permanent structures.

12.5.2 Permanent structures

Wherever possible gully control should be achieved by vegetative methods or a combination of vegetation and cheap simple structures whose life is not important. However, there are cases where the problem can only be solved by the construction of permanent structures and the main point is that for such works to be successful they must be done thoroughly and carefully. Everything is against their being successful. They will be built in adverse conditions, in poor unstable soils, in remote inaccessible areas where maintenance will be poor, and then be expected to withstand the onslaught of torrential floods and to last for ever. The gullies of the world are littered with the remains of ruined structures which

Plate 12.4 A multiple-arch silt-trapping dam in the Republic of South Africa

demonstrate that half-measures or jobs done on the cheap are a waste of time and effort in gully control.

Silt-trap dams

One example where the problem can best be solved by a permanent structure is the case of an excessive sediment load which threatens down-stream water supplies. Trapping the silt in sufficient quantity by vegetative means may be slow and uncertain. A quick positive reduction in sediment movement can be achieved by building permanent silt trapping dams. A programme of this nature was carried out in the Republic of South Africa when serious gully erosion in the Tarka Conservation Area threatened Lake Arthur. Permanent silt trap dams of many shapes and sizes were built wherever suitable sites occurred, and plate 12.4 shows an ingenious and original design of multiple-arch dam on one of the larger rivers.

Regulating dams

Another useful application of permanent dams is to regulate flash floods by what is sometimes called *the leaky bath-tub principle*. A permanent dam is built at the top of the valley with sufficient storage for the runoff from a single storm.

The outlet consists of a permanently open pipe of about 15 or 20 cm diameter which allows the flood water to drain away in a day or two, leaving the storage reservoir empty for the next storm. The flow down the gully now being reduced to the flow through the outlet pipe, it is fairly easy to create stable conditions which can cope with this flow. Practical construction points are to have the inlet to the pipe raised above the bottom of the dam outlet into an energy-dissipating chamber, not directly into the bed of the gully.

Gully-head dams

A third example of a case for permanent structures is when an active gully head is eating its way steadily upstream and must be stopped before it threatens a road or bridge or similar asset. An effective way of controlling the erosive force of the runoff over the gully head is to submerge the head of the gully in the pond of a permanently impounding dam as shown in figure 12.7. The energy of the inrushing water is then dissipated as it flows into the pond. There is one danger to be guarded against, and that is that although there is no danger of the head cutting back when the dam is full, it may do so when the dam is empty and the runoff then runs over the edge of the gully head. In a climate with marked wet and dry seasons it is possible for the head to move back a little each year before the dam fills until eventually the head is at the full supply level of the dam and so is no longer submerged and continues to eat back unchecked. This is called *the gully climbing back out of the dam* and avoiding action is to allow plenty of freeboard between the gully head and full supply level, and to adopt slowing down measures if the gully head continues to move back.

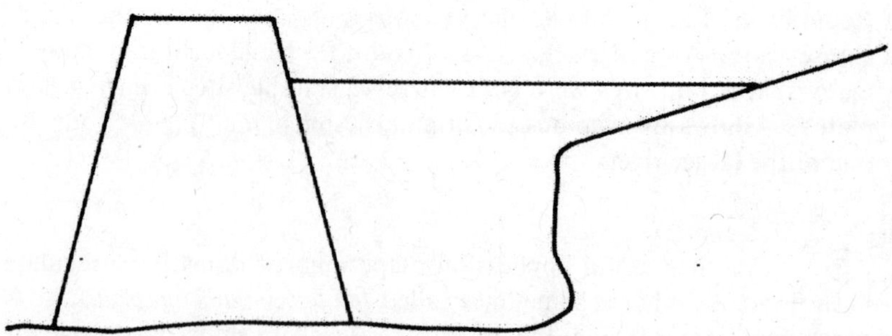

Figure 12.7 Submerging the head of the gully by a dam

Drop structures

The other approach to this problem is to stabilize the head of the gully with some masonry, brick, or concrete structure which allows the flood runoff to pass over harmlessly. A typical concrete drop structure is shown diagrammatically in figure 12.8. The capacity of drop structures of this type is controlled by the size of the inlet, which acts as a rectangular weir with the flow proportional to the length of the weir. Mathematically

$$Q = CLH^{\frac{3}{2}}$$

where

Q is the flow
L is the length of the weir
H is the head of water flowing over the weir and
C is a constant depending on the entrance conditions.

The length of the weir (L) can be increased by the addition of a box inlet as shown in figure 12.9, thus greatly increasing the capacity at little extra cost and without increasing the overall width of the structure.

This kind of structure is more likely to fail through being undermined or bypassed than through the physical failure of the wall collapsing. Particular care must therefore be given to the following points in the design and installation.

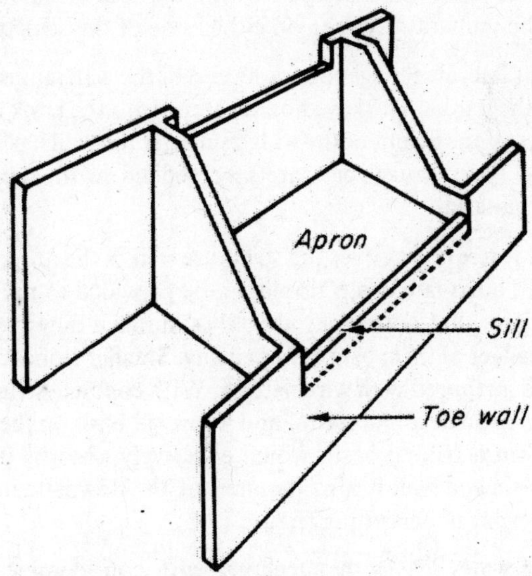

Figure 12.8 A typical concrete drop structure for gully control

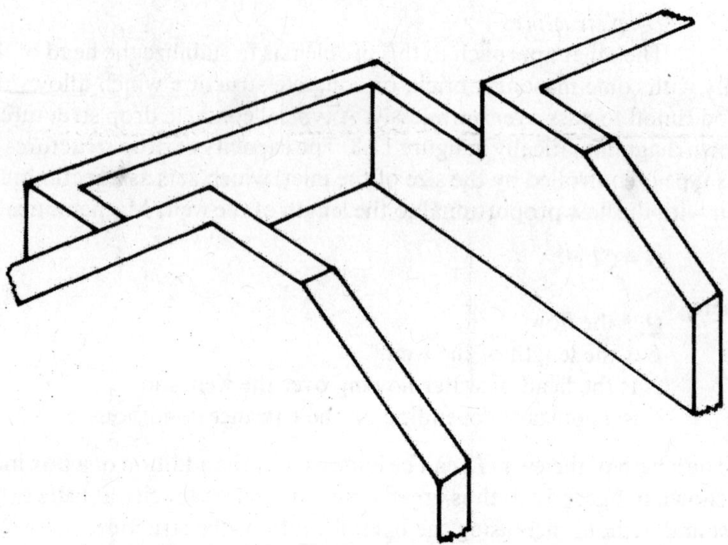

Figure 12.9 The box-drop inlet increases the capacity of a structure

1 The notch must be large enough to take the biggest probable flood. If it is too small the flood water will pass over the whole width of the structure and erode the vulnerable banks on either side of the structure.

2 To prevent lateral seepage round the end of the wall it must be carefully keyed in to the banks. The key should extend into the bank on either side for a distance of the height of the wall, and after the wall has been built the surrounding space must be carefully refilled with tightly rammed impervious material.

3 The flood water falling over the structure strikes the ground with much force, and an adequate protection must be provided to prevent scour at the foot of the wall. Loose rocks are only suitable if they are so large that there is no danger of them being swept away. Smaller stones can be packed together and anchored with wire netting. With concrete structures a good method is to provide a flat apron, and a low sill built on the downstream edge will form a stilling basin which effectively absorbs the energy. In erodible soils a toe wall is also required at the downstream edge of the apron to prevent undercutting (figure 12.8).

4 The flood water leaves the structure with considerable velocity and turbulence, so the gully sides immediately downstream may be subjected

to more severe attack than before the structure was built. Some mechanical or vegetative protection of the banks is usually required, and should be continued downstream to the point where the channel flow is normal again.

Many simpler and cheaper devices have been tried as alternatives to the drop inlet structure. Sometimes the same design has been followed but built more cheaply using timber or brickwork. Sometimes other designs have been used such as a sloping chute to carry the water down an inclined channel, and every conceivable lining material has been used: grass, asphalt, concrete, precast sections, glass-fibre and so on. No doubt some of these have been successful or partly successful, but if it were possible to determine the number of cheap or unorthodox structures which were operating successfully after a life of say 10 years, this would be a disappointing figure. A well-designed and carefully constructed drop structure will be expensive, but it will be effective, and it will last, so it will be a better choice than one which is cheap but fails after a few years.

Constructional details of some of the designs proven after long service in the United States of America are given in the standard textbooks such as SCHWAB *et al* (1993).

Gabions

The main difficulty with rigid structures is that they cannot adapt when changes occur in the soil surrounding them or supporting them. Even slight movements, such as swelling and shrinking of the soil or a small settlement of the structure, can introduce stresses which the structure is ill-equipped to withstand. Concrete, masonry, and brickwork all have good resistance to compression but fail easily under tensile loads resulting from settlement. A construction method which overcomes this problem is a more sophisticated version of the wire netting bolsters described among the temporary measures. The method was developed in Italy and uses prefabricated rectangular baskets called *gabions* made of heavy duty wire netting (plate 12.5). The basket is placed in position and filled with stones, then the lid is wired down. The baskets, up to 4 m long, 1 m wide, and 1 m deep, are built up on top of each other like courses of brickwork, and can form large or small structures. The use of galvanized wire ensures a long-lasting resistance to corrosion. The main advantage of these structures is that there is sufficient flexibility for the structure to adjust to settlement resulting from scouring of the foundations without any loss of strength. Prefabricated factory-made gabions use heavy duty galvanized wire and are long-lasting but expensive. Cheaper locally-

Plate 12.5 Gabions made on site from galvanized fencing wire in India (T. F. SHAXSON)

woven substitutes made up on site have been used in India (plate 12.6) and in Mali (RANDS 1992).

References

BENNETT, H. H. 1939 *Soil Conservation*, chapter XVII, McGraw-Hill, New York

DURBACH, S. 1964 Some simple methods of gully control, *Rhodesian Agricultural Journal*, 61, 2, 31–37

HARRIS, W. S. and R. C. HAY 1963 Gully control without structures, Part 1, *Transactions of the American Society of Agricultural Engineers*, 6, 37–39

HEEDE B. H. 1977 Gully control structures and systems, in Guidelines for watershed management, *Conservation Guide 1*, FAO, Rome

HEEDE, B. H. 1987 Opportunities and limits of erosion control in stream and gully systems, in *Erosion Control – you're gambling without it*, Proceedings of 18th Annual Conference, IECA, 205–209, Pinole, California

HUDSON, N. W. 1963 Gully control in *mopani* soils, *Rhodesia Agriculture Journal*, 60, 1, 22–31

Plate 12.6 A successful silt-trap gully control structure using home-made gabions in India (T. F. SHAXSON)

IMESON, A. 1980 Gully types and gully prediction, *KNAG Geografisch Tijschrift*, XIV, 5, 430–441

PIEST, R. F., J. M. BRADFORD and R. G. SPOMER 1975 Mechanisms of erosion and sediment movement from gullies, in *Present and Prospective Technology for Predicting Sediment Yields and Sources*, ARS-S-40, 162–176

RANDS, B. C. 1992 Soil conservation in North eastern Mali, in *Soil Conservation for Survival*, edited by H. Hurni and K. Tato, Soil and Water Conservation Society, Ankeny, Iowa

SCHWAB G. O., W. ELLIOTT, D. FANGMEIER and R. K. FREVERT 1993 *Soil and Water Conservation Engineering*, 4th edition, Wiley, New York

STOCKING M. A. 1980 Examination of the factors controlling gully growth, in *Assessment of Erosion*, edited by M. de Boodt and D. Gabriels, 505–521, Wiley, Chichester, Sussex

13

Erosion control on non-arable land

Damage to arable land tends to attract most attention but degradation of non-arable land can also have serious effects, whether economic, or social, or on the food supply. In dry seasons, loss of grazing capacity is a common and serious problem, forest land may suffer through over-exploitation, roads can cause much damage particularly on steep land. In this chapter we will look at the more important erosion problems on non-arable land.

13.1 Erosion on grazing land

On well-managed high-yielding pasture it is unlikely that there will be any erosion, for the ground will have a good uniform cover. But land which is not arable because it is too poor, too infertile, too stony, or too arid, and is only described as grazing land because it is unfit for more intensive use, is a very different matter. Again, erosion control can be directly and precisely equated to improved management – anything leading to a better water balance and a better ground cover will lead to reduced erosion.

13.1.1 Overgrazing and erosion

Overgrazing may be the result of too many livestock on the land, or the stock being on the same land for too long (plate 13.1). The management of both the numbers and the land are very different on large commercial livestock operations, or for the small-scale mixed farmer, or for the nomadic pasturalist. But in all cases getting the numbers right is important. A recent review concludes: 'Stocking rate is and always will be the major factor affecting the degradation of range land resources. No grazing system can counteract the negative impacts of overstocking on a long-term basis' (PIEPER and HEITSCHMIDT 1988).

Plate 13.1 Overgrazing results in the destruction of the protective vegetative cover

Data on livestock numbers must be treated with caution. One cannot imagine that small developing countries put much effort into collecting data on the numbers of different types of livestock. Also, in the case of cattle, there is often a marked reluctance to supply accurate information because this is equivalent to asking a businessman for details of his bank account. However, from the available evidence there seems little doubt that in many developing countries the livestock population is rising at least as fast as the human population, and in some countries at a faster rate (WORLD RESOURCES INSTITUTE 1992).

The reasons for the importance of cattle and the consequent desire to own as many as possible are well documented. Social factors include the prestige attached to numbers of livestock, their importance in marriage ceremonies, and in India their religious implications. Equally important is the straightforward financial case for owning livestock. A study in Swaziland compared the financial reward from investing in cattle with investing in other securities including land, property, and other possible investments. The answer was very clear – cattle were by far the best investment. In addition to the inflationary increase of value per head, the numbers increase like compound interest or the 'geometric progression' of Malthus.

Arid and semi-arid areas are most vulnerable to damage by overgrazing. For one thing low rainfall means low production of vegetation, and

Plate 13.2 The effect of excluding grazing, Ethiopia, with complete exclusion on the right bank and continued overgrazing on the left bank

this is compounded by the greater unreliability of rain, i.e. the greater risk of seasonal drought in regions with low mean annual rainfall (FAO 1981 figure 2.1).

The debate about whether long-term climatic changes are causing increased desertification will be long and acrimonious (for a brief summary see HUDSON 1992, page 26), but there can be no doubt about overgrazing causing significant change over vast areas of Africa because this is being recorded by satellite imagery. Two classic examples are reproduced in FAO 1987. One is from Niger, where the Ekrafane ranch, with controlled grazing, is surrounded by land denuded of vegetation (page v). What the satellite is recording is the difference in *albedo*, i.e. the reflection of light from the earth's surface. Bare soil reflects more light and looks pale, while vegetation absorbs light and shows up dark. The other example (page 130) is from Namibia where a well-run commercial cattle ranch, and the protected Etosha National Game Park, contrast with surrounding overgrazed land.

Restoration of degraded rangeland Where grazing land has been degraded by overgrazing, the capacity for it to recover when rested is sometimes quite remarkable. Plate 13.2 shows an example from Ethiopia, where the total exclusion of all livestock on one side of the river gave a rapid recovery. Another example comes from India, shown in plate 13.3. Total exclusion was enforced by a watchman, and trees for fodder and fuel were planted in a programme operated by the local community. Again, there was a rapid recovery after only a few seasons.

13.1.2 Management of grazing land

There are no simple rules about what constitutes 'the best management' of grazing land because local conditions and the management objectives will vary greatly. The principles of management for control of erosion are simple and universal. The task is to reverse the downward spiral of poor vegetation leading to high rates of soil loss and water runoff, so that vegetation becomes worse, and to substitute instead a favourable water balance with more rain being available for plant growth which will then become more vigorous. The botanical issues are also fairly clear. Over-stocking leads to a degeneration of the vegetation, with the annual grasses increasing at the expense of the perennials, and coarse tough weeds replacing the grasses, and the invasion of unpalatable shrubs, bushes and trees.

What is less understood, and currently the subject of much debate, is the effect of alternative management techniques and management systems. The mainstream range management theory assumes that there is a quantifiable carrying capacity or maximum stocking rate with several alternative definitions, usually incorporating the concept of sustained yield. *Ecological carrying capacity* is defined as the point at which livestock populations cease to grow because limited feed supplies produce death rates equal to birth rates. Most livestock managers find it profitable to hold their livestock populations somewhere short of this ecological ceiling, but what constitutes the *economically optimal stocking rate* will vary according to the producers' husbandry practices and management objectives. As pointed out by BEHNKE and SCOONES (1991) the conventional definition is based on research carried out in the United States from about 1940 to 1960 and assumes that the management objective is to maximize sustained income from the sale of beef animals, but this is not at all relevant to either small-holder farmers or nomadic pasturalists.

To illustrate the complexity of management objectives, Behnke and Scoones describe possible objectives for what might be thought to be a very simple situation, the management of a game park in Africa. The first scenario is a park which can be operated most profitably on the basis of selling permits for trophy hunting. In this instance, the park manager will want a grazing system which produces magnificent trophy specimens,

(*a*)

(*b*)

(*c*)

Plate 13.3 Restoration of grazing land, India (T. F. SHAXSON)
(*a*) the communal grazing land is almost denuded by overgrazing
(*b*) the village council agrees a policy of total exclusion of grazing animals, and tree planting, and appoint a watchman
(*c*) the trees prosper and the grass recovers after only two seasons

which means a low density of very healthy animals, what might be termed a *trophy carrying capacity*, and a by-product of this management would be abundant vegetation.

A second scenario is a park run to produce a maximum weight of game meat for sale. In this case the manager will require the density of animals which provides the maximum sustained yield in terms of meat output with more animals and less vegetation than the trophy hunting park.

The third scenario is a park which is financially sustained by a tourist industry based on game viewing. In this case the manager will require a relatively dense population of animals to increase the probability that the individual tourist will actually confront the animals he is interested in viewing. In this instance the park manager may desire a high animal population well above economic carrying capacity, a density which might be termed *camera-carrying capacity*. The by-product of this form of management will be a lower standing crop of plants than in either of the previous two scenarios.

This example shows how management objectives dictate the managerial operation. The management objectives will be very different and probably even more complex for the smallholder who is farming a mixture of livestock and crop production. In some countries there is a predominant social objective to own as many head of livestock as possible with little regard for their condition or saleable value. The desired output, that is the product of the herd, may also be quite different, with sale or meat production being of minor importance compared with the live-animal products such as milk, manure, traction power, and transport.

Going back to the question of range management techniques, more than 30 years ago research in Kenya showed that reasonable levels of productivity can be sustained on very poor quality grazing land under intensive management including rotational grazing supplemented by the storage and feeding of hay, but among small-holder farmers or pasturalists problems of land tenure may inhibit this type of management (PEREIRA *et al* 1961). Since then, many systems of range management have been developed and tested using the principle of paddock fencing which allows easily managed rotational grazing. Recently the concept of 'short-duration stocking' has been advanced, based on the theory that heavy stocking rates will not cause permanent damage to the vegetation if they are only maintained for a short time within a rotational grazing scheme using many paddocks (SAVORY 1988). However, this approach has been challenged and found wanting, for example by PIEPER and HEITSCHMIDT (1988), BRYANT *et al* (1989), TAYLOR (1989).

One result of the concept of sustainable carrying capacity has been the introduction of programmes directed specifically towards destocking either voluntarily or by decree. A typical justification for this approach is given in the case of Swaziland by FOWLER (1981) who shows that: 'The technical efficiency of the Swaziland national herd has declined at an alarming rate since the mid-nineteen sixties, and the trend shows no sign of slowing down. It cannot be explained by a worsening rainfall pattern and it would therefore appear to be the direct result of pasture degradation by overstocking. The unpalatable conclusion appears to be that some form of control of livestock numbers is essential.' However, destocking programmes have always been unpopular and as a result few have been successful. An interesting example is the HADO project in Tanzania where complete destocking of a badly degraded area was apparently accepted by the local population when pursued vigorously by a strong socialist government (MNDEME 1992), but the long-term implications are now being studied (CHRISTIANSSON *et al* 1991). A similar programme of total exclusion of livestock for a limited period of time was practised in Ethiopia (CHADHOKAR 1992), but that too was enforced by an authoritarian socialist regime which has now been overthrown so its future is uncertain.

One reason for the reluctance to accept the western ideas of carrying capacity and stock rates is that many livestock managers in areas of variable rainfall practice what is called *opportunistic stocking*. This is maintaining herds large enough to make full use of vegetative growth in seasons of good rainfall although this carries the risk of a bigger loss of stock in drought years. This is often the strategy of nomadic pasturalists who practise seasonal migration. The disadvantage of this approach is that drought reduces the herd quickly, but it takes some years to build up again. The herd size therefore responds to and follows variations of rainfall, but with a time-lag. A suggested improvement is a 'tracking system' which provides a quicker response to the variations in rainfall (SANDFORD 1983). Among the techniques used in this system are

- 'The establishment of locally managed grazing territories in which a management fee per animal is charged. The fee is inversely related to the previous season's rainfall, so that it is expensive to keep cattle after a dry season, and cheap after a wet one.
- 'A subsidized market price, also inversely related to rainfall, so that after a dry season the price is high, in a good year low.
- 'Improved marketing, trekking, and abbatoir facilities, so that stock can be removed from the range and slaughtered quickly.
- 'Improved drought recovery measures' (ABEL and BLAIKIE 1989).

Overstocking and degradation can arise, particularly in developing countries, when communally-held rangeland is grazed by privately-owned stock. Individual owners who increase their herds gain the marginal benefit, but the marginal cost of land degradation and reduced vegetation is borne by the whole community – a situation described as 'the tragedy of the commons' (HARDIN 1968).

The solution appears to lie in making grazing rights a saleable commodity like land freehold, or in transferring management and responsibility to groups of users, or a combination of both within some framework of self-management. Many such schemes are being developed and tested in Africa, such as the government-sponsored but democratically managed Grazing Associations in Lesotho (HUNTER and WEAVER 1993), and the government-led Arid and Semi-Arid Lands programme (ASAL) in Kenya (BIAMAH 1992). Other examples are reported from Niger by PACEY (1986) and from eastern Senegal (UNDP 1984). Programmes like these, with a strong element of local participation, are more promising than the programmes of area closure enforced by decree through the local branch of the ruling political party.

13.1.3 Physical erosion controls on grazing land

The usual objective of physical manipulation of grazing land is to improve moisture availability by increasing infiltration (for example by ripping as discussed in section 13.4.2) or by temporarily catching some of the rainfall in pits or drains. Some of the practices used on arable land and described in sections 10.2.2. and 10.3.2 are also applicable on grazing land.

Mechanized practices

Managers of grazing land (rangeland in the USA) are often dealing with large areas, so there is an opportunity to use practices which are put in place by either conventional farm machinery or special purpose-built equipment.

Pasture furrows are small open drains on level grade following the true contour and fairly close together. The individual furrows are small, often made with one pass of a single-furrow plough and closely spaced only a few metres apart. Sometimes the soil from the furrow is spread entirely on the uphill side so that there is no bank. In this case a furrow which is catching more water than it can hold helps to distribute the water by spilling it over the whole length of the level downhill edge of the furrow.

Another approach is to make many small surface depressions which hold and store runoff. *Range pitting* is a simple one-pass operation pulling a modified disc plough which intermittently scoops out shallow pits as shown in plate 13.4 and figure 13.1. A recent variation on this theme is *imprinting* developed in the USA and illustrated in plate 13.5 (PRATT 1983, DIXON 1987). A number of other special-purpose machines to create similar effects have been developed for the dry regions of western Australia and are described in FAO *Soils Bulletin 57* along with other aspects of water management in low rainfall conditions.

Permeable barriers may also be used on non-arable land where the main objective is to temporarily detain surface runoff and trap silt where the increased soil moisture can encourage improved grass growth. The semi-circular hoops and trapezoidal bunds described in section 10.2.2 as an improvement of arable land may also be used on grazing land, where they may be simple low stone walls as in plates 13.6 and 13.7 where the walls are built with lumps of laterite. The hoops may be connected together to form continuous lines as in plate 13.8. In Burkino Faso and Mali the use of

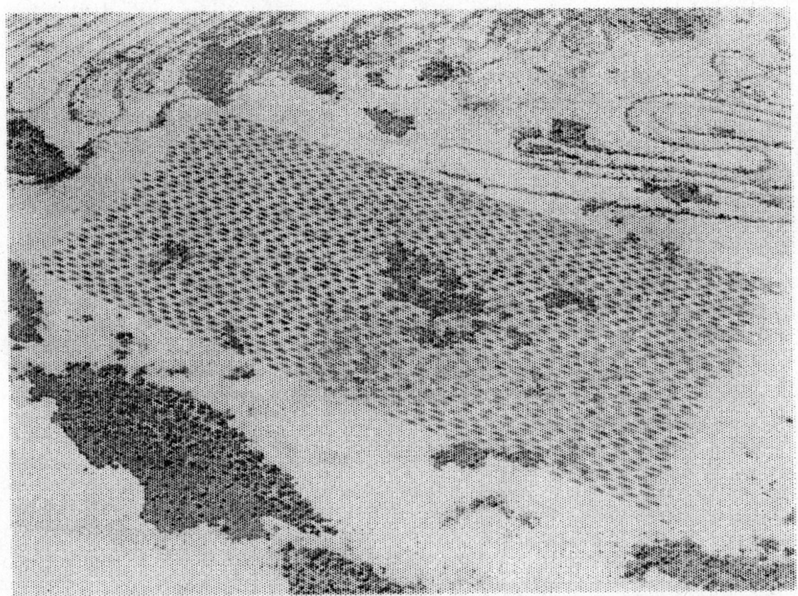

Plate 13.4 Range pitting in the semi-arid western block of New South Wales, Australia (Western Australia Department of Agriculture)

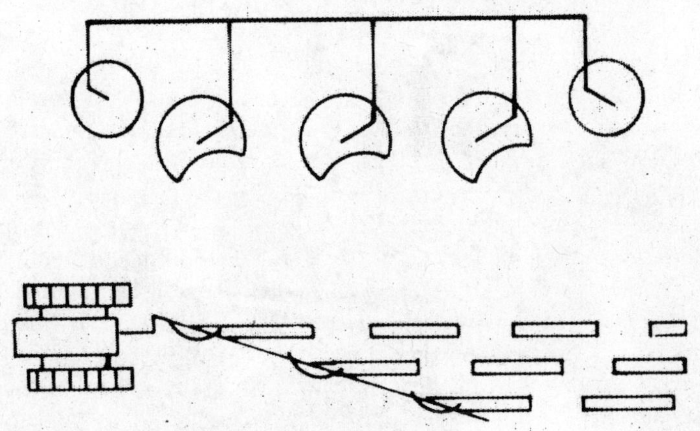

Figure 13.1 Modified disc plough for range pitting

Plate 13.5 The *imprinter* improves soil surface storage of rainfall by creating a corrugated surface (R. M. DIXON, USDA–ARS)

Plate 13.6 Marking out semi-circular hoops called 'diguettes' with a shallow excavation, Burkino Faso (OXFAM)

Plate 13.7 Building the low stone wall with lumps of laterite, Burkino Faso, (OXFAM)

Plate 13.8 Stone lines set out on the contour near Onahigoua, Burkino Faso (OXFAM)

stone lines in this manner has the primary objective of water harvesting rather than soil conservation. Runoff from uncropped land higher up the slope runs down onto the crop land and is spread by the permeable stone lines.

13.2 Erosion control on roads and construction sites

13.2.1 Siting and alignment of roads

The surprising thing about the problems connected with farm roads is how many of the problems could have been avoided. In developed countries the siting of roads is usually determined by existing features such as boundaries, but when putting in roads for the first time in a new agricultural development scheme, getting the roads right is one of the key steps (MEGAHAN 1977, FAO 1989). The siting of new roads is one of the many tasks which nowadays can be done so much more efficiently by the use of aerial photography. Using a stereoscope to study pairs of photographs the planner sees a three-dimensional model of the ground on which crests, valleys, and other topographic features can be quickly and accurately picked out. Even without using the stereoscopic principle to get a 3-D image, the aerial photograph greatly simplifies the siting of new roads. The most striking example is planning the line of a new road through rugged country with indifferent maps and heavy vegetation. A ground party laboriously hacking its way through on foot and by Land Rover, frequently wasting time by back-tracking from false starts and detouring round obstacles, can take months to do what could be done in a day using aerial photographs and the simplest interpretation equipment.

The first rule of road siting is to put roads on crests wherever possible, thus disposing of one of the most troublesome problems – drainage. A road along a crest has no catchment to shed water onto the road, and the runoff from the road surface can be simply discharged on both sides. There is no need for any bridges, culverts, or crossings on a road crest, and maintenance is simpler and required less frequently. This applies equally to all kinds of roads, from main road to the farm track through arable fields.

When it is not possible to put the road on a crest, the next best alignment is on a gentle grade fairly close to the true contour. Gradients of the order of 1 in 100 to 1 in 500 will present no difficulty to traffic and are best for the open-channel drains which will be required alongside the road. Grades from 1 in 100 to 1 in 20 are not difficult for traffic, but controlling the erosion in the side drains may be more of a problem.

The worst choice is the road which goes diagonally down the side of a hill on a grade steeper than 1 in 20, and it is usually better to use a zigzag layout, or the combination of some lengths on gentle grades and some parts straight down the slope, as in figure 13.2.

13.2.2 Road drainage

In all soils the strength or resistance to deformation reduces greatly when the soil is wet, so a road cannot work efficiently unless it is

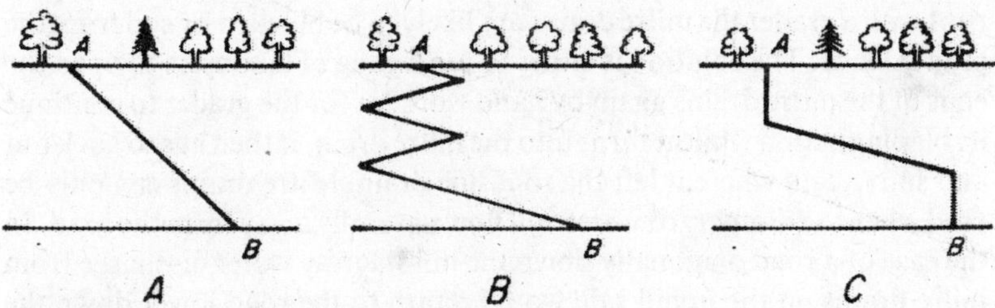

Figure 13.2 Road alignment down steep hills
A is undesirable because drains on so steep a grade must be lined and are expensive
B or *C* using grades suitable for grass-lined rain channels are preferable

properly drained. When siting roads, swamps and permanently wet areas should always be avoided for this reason. Getting the water off the surface of the road is fairly easy, and on a new road surface all that is required is a gentle camber. On earth or gravel roads, wheel ruts may prevent the water getting to the side of the road, and on a gradient there is a danger of each rut becoming a scouring watercourse. The solution is to divert the water sideways by very gentle depressions on the road surface. If made sufficiently wide these cross-ditches can be negotiated without any difficulty by traffic and by maintenance equipment such as graders. The size of the drain depends on the gradient and the speed of the traffic. On a farm road carrying only tractors the cross-drain need only be the depth of the wheel ruts and perhaps a metre across, whereas on a gravel road carrying fast-moving cars a gently dished section is required, with a width of up to 20 metres.

Drainage along the side of the road is usually the more difficult problem, not so much because of the runoff from the road itself, but because the road is cutting across natural drainage lines and so picking up runoff from any nearby higher land. Only crest roads are entirely free from this problem and can manage without any side drains.

Roads straight up and down the steepest slope need side drains only to deal with the runoff from the road surface, and this water can be easily disposed of by *mitre drains*. These are extensions of the road drains leading away from the road at an angle of about 45° so that the storm runoff is dispersed before it can build up to unmanageable quantities and velocities. The spacing of mitre drains will depend on the intensity of rain expected, but would usually be from 20 metres to 100 metres. The cross-section must be at least as large as the drain alongside the road, for any restriction will cause a decrease in velocity and the deposition of sediment. When the road surface and the side drains are maintained by smoothing runs with a grader the mitre drains are likely to be blocked by soil from the grader blade. The solution is either to use a gang of labourers to open the ends of the mitre drains again by hand work, or for the grader to continue its blading action while it turns into the mitre drain. It then has to back out and start again where it left the roadside drain. Mitre drains can only be used when the discharged water will flow naturally away from the road. In the case of a road diagonally down the hill side any water discharge from mitre drains on the uphill side would return to the road lower down the slope.

For roads on gentle gradients open side drains are required only on the upper side, and these should be designed on the same principles as vegetated stormwater drains as discussed in section 10.4.2. There is no point in having a drain on the lower side, for any unnecessary collection and concentration is undesirable.

The main reason for avoiding roads running diagonally down a hillside is the difficulty of draining them. It is easy to design and maintain vegetation-lined open ditches on gradients up to about 1 in 50, but anything steeper is not suitable for earth channels. This means either lining the drains or passing the water under the road at frequent intervals by culverts. Both these methods add greatly to the cost of simple low-cost roads.

If a design specification was requested for open drains at the sides of roads it could be very simply expressed as 'make them mowable'. A wide shallow cross-section with gently sloping sides will provide the best hydraulic design, and regular mowing of an established close-growing cover grass has been shown to be the most effective and the cheapest maintenance (plate 13.9).

When simple earth or gravel roads are maintained by periodic grading with a self-propelled road grader problems can arise from the shape of drains which develop. The grader takes a small slice from the nearly vertical face of the drain and spreads it on the road, leaving a V-shaped ditch which is the shape most vulnerable to erosion (figure 13.3). Flat-bottomed drains are preferable, especially if a suitable grass can be established and then maintained by mowing. Even if the drain is kept clear of vegetation by an out-of-date road engineer who insists on bare drains, the risk of erosion is still reduced.

Mechanical structures in the drains are highly undesirable. They are expensive to install, they require regular care and maintenance, and worst of all they seldom do their intended job of stopping erosion. With good choice of alignment, and properly designed open drains they can usually be avoided. The exception is at the entrance to culverts taking runoff under the road, where drop-inlet structures are usually desirable (figure 13.4). They give better entry conditions to the culverts, perhaps allowing the use of a smaller pipe, they lower the effective gradient in the drain, and they reduce the depth of excavation necessary in the drain.

13.2.3 Stabilization of banks and cuttings

Steeply sloping earth banks are liable to landslides and the key to the control of this form of erosion lies with drainage. Wet soil has both a greater weight and a reduced resistance to deformation, so a bank which is stable when dry can become unstable when wet. As with all drainage, the first principle is to intercept and divert as much water as possible before it arrives at the point where it becomes a nuisance. When a road has been

Plate 13.9 Good modern road drainage design using a woven fibre netting to provide protection until the grass is established (BEMIS CO.)

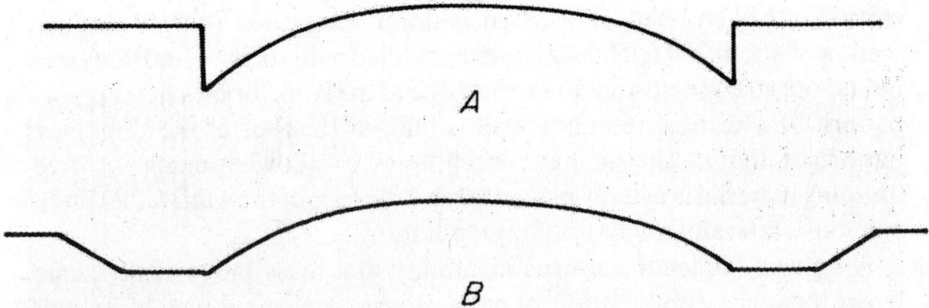

Figure 13.3 The shape of graded road drains
A with V-shaped drains is undesirable
B with mowable drains is preferable

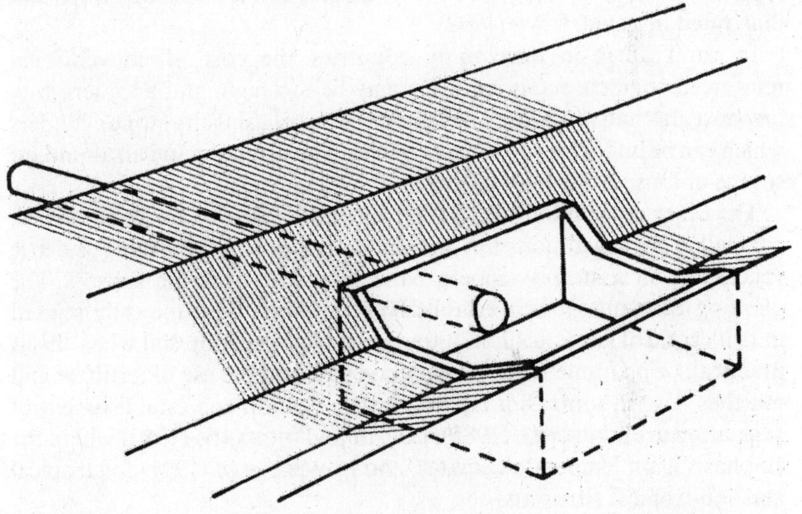

Figure 13.4 Drop inlets improve the discharge efficiency of road culverts

cut into the side of a hill, producing a steep bank on the hill side of the road, it is most probable that above the bank there is more hillside from which surface runoff comes down towards the road. This should be diverted by storm drains, designed as the diversion drains to protect arable land (section 10.4.2). The diversion drain should be sited at some distance uphill from the edge of the embankment. If the drain is close to the edge, there is the danger that when flowing full it will add to the weight and also seepage from the drain may aggravate the very problem it is trying to solve. Drainage of the sloping face of the embankment is usually carried out by rubble drains, that is drains excavated to a rectangular section of about 300 to 500 mm^2 and backfilled with rocks. The drains may be laid out straight up and down the slope of the bank, or on a herringbone pattern. If a vertical retaining wall is built at the foot of the slope, it is important that it should have weep-holes to allow drainage. A free-draining material is usually placed behind the foot of the wall so that water can move laterally along to the weep-holes.

A natural tendency towards instability and mass movements occurs when there is a combination of steep slopes, geological instability, and high rainfall. Examples are Nepal, Hong Kong, and Japan, and in each of these much effort has been devoted to research on control methods – mainly high-tech methods for high-value urban sites in Hong Kong and Japan (see *Further reading*) and low-cost solutions in Nepal described and illustrated in NEPAL SPWP (1992).

In rural areas in developing countries the cost of conventional reinforced concrete retaining walls may be too high, and an interesting low-cost alternative has been developed in Nepal using dry stone cylinders which can be built cheaply because of cheap local labour and an abundant source of loose stone (LAWRANCE *et al* 1991).

The other erosion hazard, that of surface erosion of the bank, is best controlled by vegetation, but the difficulty lies in establishing a dense vegetation on a steeply sloping bank of usually infertile subsoil. The planting techniques to help establish vegetation are the same as those used in gully control (section 12.4.2), that is replacing the top soil when this is practical, or planting in pockets of good soil, and the use of fertilizer and mulches. Useful sources of advice on the selection and establishment of vegetation are SCHIECHTL (1985), COPPIN and RICHARDS (1990) where the emphasis is on temperate climates, and HOWELL *et al* (1991) for tropical and sub-tropical situations.

On gentle slopes mulching materials spread on the surface may be sufficient to assist germination and reduce runoff and erosion while the vegetation is being established. On steeper slopes the mulch may have to be fastened to the surface to hold the mulch in place or to serve the same purpose as a mulch. Straw and hay are the most commonly used mulching materials, but any organic material will serve provided it can be finely divided and evenly spread. In the United States successful use has been made of *Excelsior* which is fine curled wood shavings originally produced as a packing material (plate 13.10). Several types of netting are manufactured specially for holding in place a loosely spread mulch. Natural materials include netting woven from twisted kraft paper yarns or from jute yarns, and in recent years a huge range of synthetic fibre

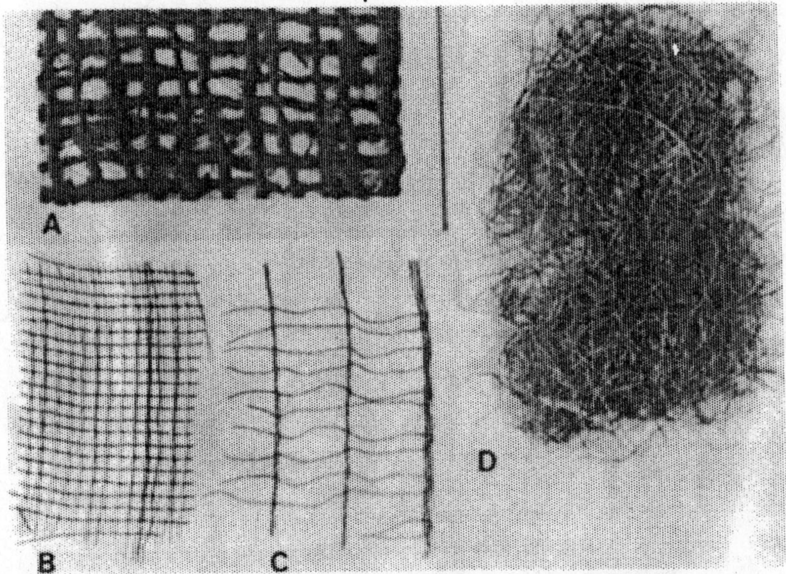

Plate 13.10 Some materials used to assist the stabilization of slopes *A jute-net, B erosionet, C mulch-net, D Excelsior*

materials have been developed. To use these, the slope is smoothed, seeded, and fertilized, then the layer of mulch is spread, and the netting unrolled over the mulch and anchored by a wire staple rather like an oversized version of those used to fasten papers together. The netting which is intended only to anchor a straw mulch is very light with a loose open weave. Others, more closely woven from coarser yarns or denser synthetic mats can themselves serve as a mulch and are stapled directly to the soil after seeding and fertilizing.

Natural materials are usually biodegradable which may be cosmetically an advantage where a natural cover of vegetation is the objective, but the longevity of plastic materials may be desirable if establishing the vegetation is going to be slow. Some examples of natural fibre mats are shown in plates 13.10 and 13.9, and synthetic materials in plate 13.11. In both cases the structure may be either woven or non-woven. Tests of the relative effects of a group of natural and synthetic geotextiles used as surface mulches showed that in general the thicker mats are more effective than thin ones (RICKSON 1990).

There are many possibilities for using vegetation strengthened by some form of physical reinforcement, such as cellular concrete walls (called *crib*

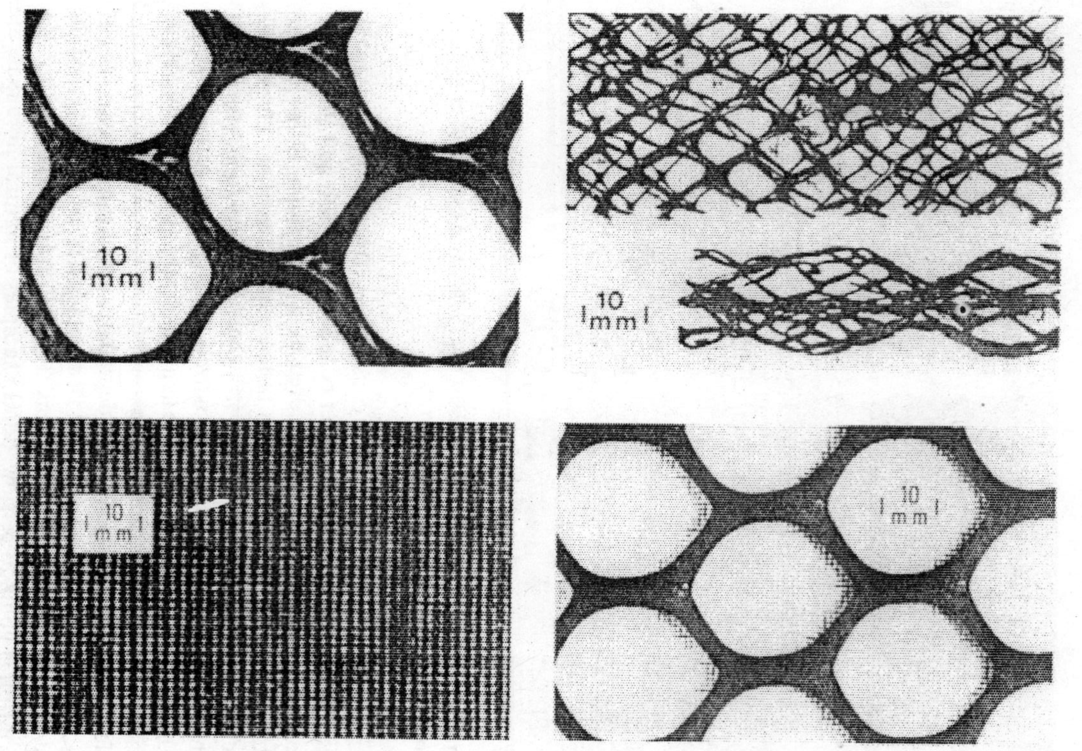

Plate 13.11 Some synthetic geo-textiles

Plate 13.12 Steep roadside cuttings may need structural support which may be combined with vegetative plantings

walls) with interplanting (plate 13.12) or contour lines of brushwood (plate 13.13). Another technique which is finding increasing favour for accelerating the establishment of vegetation on slopes is *hydraulic mulching*. A mixture of water, grass seed, fertilizer, and a filler is sprayed through a high-pressure hose to form a thin film on the soil surface. One material used as the filler is finely-chopped cellulose fibres, and this has the effect of both forming a thin protective mulch against rain erosion, and also providing enough moisture for the seed to germinate if no rain falls for a few days after application. Since the mulch is a layer of unconnected particles it is not resistant to surface flow and so this technique is only suitable for gentle slopes, but it is increasingly popular for road reservations or housing sites where a quick cover justifies the cost. A small shoulder-carried sprayer was used successfully to establish grass on roadside cuttings in Jamaica by SHENG (1978). The method is also popular on the steep slopes of railway cuttings or steep road embankments where the ability to spray using heavy equipment on the road or railway track allows large areas to be covered quickly. Mineral fillers based on gypsum, or perlite, or vermiculite are also used.

Alternatively, instead of a filler to bulk up the mixture, there may be added a water-based emulsion, and examples in use are oil, bitumen, latex,

Plate 13.13 Pegged lines of brushwood can be temporarily effective to assist the establishment of vegetation

and gums, known in the USA as *tackifiers*. This kind of mixture sprays a thin surface layer which in effect sticks down the soil's particles while not restricting germination, indeed often improving it. For some of these emulsion mixtures the recommended procedure is first to prepare a seed bed, apply the seed and fertilizer and then spray the mixture on top. In other cases the seed and fertilizer may be added to the spray mixture. Another variation is to combine the bulk of a straw mulch with the fixing power of bitumen. Two nozzles are used, one spraying chopped straw blown by a high-pressure air supply, the other spraying a bitumen emulsion in water. The streams of material from the two nozzles intermingle before landing on the soil surface. The spray-on methods are all fairly expensive because of the materials and the high labour and equipment costs. They therefore find little application in the control of erosion on farm land, but are increasingly used for the control of roadside erosion where the cost is justified by the need to establish vegetation quickly (GOLDMAN *et al* 1986).

A series of tests was carried out in a semi-arid environment in Colorado to assess the effectiveness of controlling the erosion and assisting the establishment of vegetation on sloping banks. The treatments were hydraulic mulching without a tackifier, straw mulches sprayed with commercial liquid polymers, and organic and synthetic mats and blankets (FIFIELD *et al* 1989). All were effective, and a choice of which to use depends mainly on availability, cost, and ease of application.

Another technique which is unlikely to be cost-effective on farm land, but may be justified on small sensitive areas, is to improve the soil's resistance to erosion by chemical additives. Most are complicated polymers marketed under trade names such as *Krilium* or *Agri-SC* (FULLEN *et al* 1993), or phosphogypsum, a fine-grained gypsum formed as a by-product of phosphate fertilizer manufacture (SHAINBERG *et al* 1990). All of these chemical additives work by increasing the aggregation of soil particles, and the effect is usually not permanent, but this is not necessarily a disadvantage when the object is to accelerate the establishment of vegetation. Most of the purpose-made chemicals are expensive (except phosphogypsum which is a by-product) so studies have been made of minimum rates of application (WALLACE and WALLACE 1986).

13.2.4 Urban erosion and construction sites

On urban construction sites there is a risk of high rates of erosion, especially when large areas of soil are exposed during construction (plate 13.14). This can be greatly reduced if appropriate techniques are planned into development programmes, and there is no need for erosion to be a problem left behind when development is complete. But developers are not always environmentally minded, and problems can arise from increased runoff and increased sediment yield. Concern about off-site damage has therefore led to specific recommendations, *eg* by the United States Environmental Protection Agency (1971), and much debate (FIFIELD 1991), and official guidelines such as those issued by the Soil Conservation Service of New South Wales (QUILTY *et al* 1979). Some States in the United States of America have passed controlling legislation such as the State Stormwater Management Law of Maryland, 1982. All such plans have been strengthened by increasing concern about water quality. The difficulty lies in getting the recommended practices applied on the ground. A detailed study was carried out in Sydney by TAI (1992) who on 68 construction sites evaluated the compliance with each of 20 practices recommended in the Soil Conservation Service guidelines. There was a large variation in the extent to which the various practices were implemented, but the assessment was that overall it was 'a very unsatisfactory situation'.

The effect of the legislation in Maryland was equally disappointing. Early attempts were made to control erosion in Montgomery County in 1965, followed by State-wide legislation in the Natural Resources Article

Plate 13.14 Excessive erosion can occur on urban construction sites if preventive measures are not used

of 1970. This demanded that an erosion and sediment control plan should be submitted to and approved by the local conservation district before any development was started, and included very stiff penalties of heavy fines and the possibility of imprisonment for violation, but this legislation had little effect because of inadequate procedures for enforcement. In 1982 the legislation was revised, and backed up with large sums of money for inspection and enforcement (BENNER 1985). It will be interesting to see if this will be more effective.

Another approach is being developed by landscape architects in the United States. The control and drainage of infrequent but large flash floods is expensive but can be reduced if the peak flow can be reduced. One method is to include in urban development schemes areas designed for high infiltration and successful examples are quoted by FERGUSON (1990). Another approach is to include in the overall design storage basins, lakes, and wetlands, which can provide temporary flood storage, and some of the cost can be charged to the aesthetic and recreational value (FERGUSON 1991). This appears to be a more promising approach to the problem of urban erosion than the enforcement of legislation.

13.3 Erosion and forests

The problem of soil erosion from land covered by woods and forests is usually less severe than the erosion from arable land. The protective canopy of vegetation is usually good, often giving complete protection, because there is likely to be an accumulation of forest litter on the soil surface. The result is a lower rate of erosion, and reduced peak floods compared with similar land under other uses. However, it does not follow that any forestry will automatically bring these results. In a study which compared the kinetic energy of falling rain with that of the through-fall under a forest canopy, the rather surprising result was that the energy was actually greater under the trees (CHAPMAN 1948). The explanation is that the rainfall was of a fairly low intensity, and so would be composed of drops of many sizes, all falling at terminal velocity. Rain is intercepted by a tree canopy, and eventually drips through, but the size of the drops which drip from the leaves is greater than the natural raindrops. The canopy was in this case about 10 metres from the ground, sufficient for the water drops to reach terminal velocity, so the net effect was that even though some of the rain was lost by evaporation and interception, and some would become stem-flow down the trunks of the trees, yet the total energy of the drips under the trees was greater than that of the rain reaching the ground direct. Chapman's results have been supported by later studies in New Zealand by MOSELEY (1982) and BRANDT (1987).

The result should not be over-emphasized, for it might be different with a different kind of tree canopy, or more intense rainfall. Nor does it follow that erosion would be high, because it is usual to find either an understorey of lower vegetation or an accumulation of leaf litter, and either would effectively protect the soil. What the result does show is that tree cover does not automatically provide protection against floods and erosion. This is particularly true in the tropics and sub-tropics. In temperate climates with slow rates of oxidation and decomposition there usually is a build up of surface litter under any permanent trees or shrubs, but in tropical and subtropical climates it can be quite different. The rate of chemical and biological breakdown is much faster, and the chance of physical removal probably higher, whether by erosion, consumption by browsing animals, or by harvester termites. Plate 13.1 shows the ground under trees stripped completely bare by a combination of these circumstances.

13.3.1 Mechanical protection of forest soils

Mechanical protection is usually not required for natural forests, but commercial plantings may well need some protection during establishment, and after harvesting. Two forms are most common, contour trenches and contour furrows, and both are similar to the works used to protect arable land.

Contour trenches are commonly used in America on steep lands from 15 to 35 degrees (CUSKELLY 1969). Large areas of the United States National Forests were protected in the 1930s by trenches dug by hand, but nowadays larger trenches are constructed using heavy earth-moving equipment. The trenches are usually built without any gradient in the channel since the object is not to lead off surface runoff but to hold it until it infiltrates. Cross-ties are added every 10 or 15 metres to further restrict water movement, but the height of the cross ties is made lower than the main bank so that overtopping would first take place along the trench. This safety device is the same as used in tie-ridging (section 10.5.2), and like tie-ridging, this protection method should only be used when it will be able to absorb all the rain and so eliminate surface runoff.

Contour furrows are similar in form, but smaller, and used on gentler slopes up to about 20 degrees. They have a smaller water-holding capacity but are quicker and cheaper to construct. Several implements have been developed for their construction (CUSKELLY 1969), some on the principle of modified large-disc ploughs, some using angle-dozers, in both cases mounted on a large crawler tractor.

A similar device but smaller and usually dug by hand is used to gain a local increase in soil moisture in fruit orchards and for other tree crops. These may be variously known as *contour ditches*, or *gradoni*, and may be continuous along the contour, or in short sections between drainage lines. Other similar variations are discussed in section 10.5.3 under Intermittent terraces.

13.3.2 Forest roads

The roads on forest lands present a considerable erosion problem for several reasons. The land is usually steep and the rainfall heavy. The roads are not used much and so expensive roads with careful construction are not justified. But the road use during harvesting operations is very damaging and bound to be associated with a high risk of erosion.

One question to be decided is at what stage in the forest development should the road system be constructed. Economics and the erosion hazard must both be considered (GARDNER 1979). If the roads are put in at planting, there is a lot of unproductive capital lying idle, and a long time for erosion to occur on the road and the banks. On the other hand to delay the construction until extraction starts, means struggling with inadequate access for fire protection and for pruning during the main life of the plantation. It also means the probability of severe erosion during extraction over new roads which have not developed any cover on the raw scars inevitably associated with road construction. The most economical time to construct the roads is thought to be just before the first thinning (HUGGARD 1958) and this is also a good solution from the point of view of minimizing the risk of erosion.

An important point is that disused roads will be a constant erosion problem if they are just abandoned, with the near certainty of continuing deterioration until they become gullies. Simple and inexpensive control measures can prevent this. Some form of drain or ditch should be cut across the road at intervals to prevent the build-up of runoff down sloping stretches of road. Vegetation should be encouraged, and the two basic steps are scarifying the surface and seeding.

13.3.3 Forest management and erosion

During the planting, establishment, and the growth of a plantation the management practices required to minimize erosion are straightforward – a foundation of mechanical protection works followed by management aimed at maintaining a complete protective vegetative cover. The problems really arise during harvesting when the cover must be disturbed and the timber must be extracted, both of these tending to cause erosion. The choice of clear felling or selective extraction usually has to be decided on economic grounds, but from the conservation point of view selective felling is much less damaging and so preferable where possible. There is less drastic disturbance of the tree canopy, and also a good chance of some of the ground level vegetation surviving.

Clear felling is necessary with some species, or may be the most economic method. The tree cover will be entirely removed and the extraction usually leaves the surface cover largely destroyed. It is often debated whether the hauling of logs is better concentrated in a few routes which will be badly damaged, or whether to disperse the routes and spread a lower rate of soil disturbance over the whole area. In fact a more important issue is not the extent of the damage, but the restoration afterwards. A steep hillside badly scarred by clear felling is liable to continue to erode for years afterwards, with the log hauls and roads turning to gullies and irreparable damage being done before a protective mantle of vegetation is re-established. Like a hospital patient after major surgery, the recovery can be greatly assisted by suitable aftercare, in this case by digging cross-drains across the roads and log hauls, by not burning off the brushwood and trimmings, and by getting the next cycle of vegetation under way as quickly as possible, whether it is replanting another tree crop or establishing grass or natural vegetation. The use of *nurse crops* is recommended to provide a quick cover during the first years after a new tree crop is planted.

As in other aspects of management we find again the principle that good husbandry practices coincide with good erosion control practice. The forester concerned with yield and quality is anxious to keep out fire, so is the conservationist, for even if the trees are not damaged the forest floor litter will be. Invasion by bush and other vegetation is another problem. The forester does not want his stand of commercial timber debased with other tree species, but does want a tidy understorey. The forester and conservationist would both agree that the ideal combination is a clean regular stand with a close ground-cover of grass or similar vegetation.

13.4 Reclamation of degraded land

Some land is destined by nature to be barren and subject to severe erosion, and it would be impractical and uneconomic to try to convert it to productive farm land. Other land has slipped into poor state through misuse or neglect and it is here that reclamation measures are justified. We can divide the problem quite arbitrarily into *rejuvenation*, that is when all that is required is a helping hand to improve a potentially stable situation (like aspirin for a healthy body temporarily out of sorts), and *restoration*, that is when there is something organically wrong which must be cured before the patient can recover.

13.4.1 Rejuvenation

To continue the medical simile, the symptoms of land requiring rejuvenation are easily recognized: sparse vegetation, high runoff, low infiltration, high rates of soil erosion. But which is the cause, and which the effect? The poor cover leads to high runoff and low infiltration, which in turn prevents better vegetation. The problem is how to turn this downward spiral of degradation into a rising spiral of better cover giving more available moisture, producing better cover. The corrective technique is to identify the limiting factor and improve it. Perhaps the problem is the rainfall/runoff relationship. There is sufficient rainfall for a better growth but too much is lost as surface runoff. Some mechanically constructed storage such as furrows, basin listing or one of the other methods discussed in section 13.1.3 may increase the available moisture enough to change the whole pattern. In Kenya a practice called *Katumani pitting* is used to rehabilitate land which was degraded by communal grazing but has now passed into private ownership. Small pits of area about 2 m^2 are dug by hand (like small versions of the semi-circular hoops of section 13.1.3) and planted with nurse crops of forage legumes for a season, followed the next season by broadcasting grass seed (GICHANGI *et al* 1992). Alternatively the basic limitation may be an inherent lack of fertility, or a chemical imbalance which inhibits plant growth. Application of chemical amendments or of manures, or of artificial fertilizer may be the solution. Or perhaps present vegetation is unable to make the best use of the ingredients for plant growth which are already available, and an improved variety or a different species could profitably be introduced. In the United States of America some dramatic improvements of degraded range land have been achieved by replacing the indigenous scrub vegetation with improved grass varieties. In New Zealand aerial seeding of a grass/legume mixture has improved degraded grassland.

13.4.2 Restoration

When the situation is so bad that major remedial measures are required, a popular line of attack is improving the infiltration either by surface manipulation or by ripping or by subsoiling. Large areas of low-value land are usually involved, so machines which can cover the ground quickly are preferable. In section 13.1.3 on range management we mentioned the use of range pitters (figure 13.1 and plate 13.4), and imprinters (plate 13.5). Other techniques are heavy spiked rollers and opposed disc pitters (FAO *Soils Bulletin 57*).

When the infiltration is restricted by a plough pan or a naturally occurring obstacle to percolation such as a layer of laterite, it may be appropriate to use deep ripping to break the obstructing layer. However, if the primary purpose is to improve infiltration at the surface, then ripping is an expensive way of achieving what might be done equally well by a cheap surface-scarifying operation with a disc harrow or spike-tooth harrow. Many reclamation experiments with sub-soiling show, after a few years, an improved vegetative growth along the lines of ripping, but little trace of the vegetation spreading onto the ground between. If, as is often the case, the problem is poor surface infiltration, then instead of ripping to perhaps half a metre in depth at a spacing of a metre or so, a better result would be achieved more efficiently and cheaply by an all-over discing operation.

The use of heavy machinery needs substantial improvement of the grassland for it to be cost-effective, so for small-scale or low-cost restoration, vegetative methods are likely to be more cost-effective. When the annual rainfall is sufficient in quantity and reliability, aforestation may be the answer. A detailed review of recommended species and methods is given by HARCHARIK and KUNKLE (1978).

Other vegetative measures may also be appropriate. A major study has been carried out by USDA of grass species suitable for restoration, and of techniques for harvesting the seed and sowing it (BERG *et al* 1992). Since land in need of restoration is often low in nitrogen, a research programme in Indonesia has concentrated on legumes. Several inedible legume species were found to be suitable, particularly *centrocema*, but the farmers preferred to grow food crops, and *mucuna* is the most promising edible lucerne (SUWARDJO *et al* 1991).

13.4.3 Special reclamation situations

Scald Another reclamation activity extensively studied in Australia is improving what are known locally as *scald* areas, where a combination of accumulated salinity and surface crusting leads to an almost barren surface. Many reclamation techniques have been tested on these areas, and one of the more successful is ponding behind low banks to leach out the salinity as reported by JONES (1967) and illustrated in plate 2.6. Other treatments are ripping or furrowing in spirals (plate 13.15), and sowing grass in pits (plate 13.16).

Mining residues Spoil dumps – the discarded residues of mineral extraction – are usually very erodible because the material has been

Plate 13.15 On level scald areas, spirals of furrows are used to encourage regrowth (Conservation Commission of the Northern Territory of Australia)

physically ground up during the extraction process, so there is no soil structure. Also in some cases there is a severe chemical imbalance which inhibits the growth of vegetation either from what is being mined, as in bauxite and copper, or because of chemicals used in the extraction process, as in cyanide used in the extraction of gold. Special techniques are required for reclaiming such soils and may include temporarily impounding water onsite to leach out undesirable chemicals, the application of neutralizing chemicals, the use of tolerant vegetative species, or covering the site with a layer of topsoil. Open-cast mining may also cause restoration problems, but developed countries are able to solve this by requiring that after extraction the site must be restored to a level of agricultural productivity as good as or better than before. The cost of stockpiling topsoil and replacing it, and reseeding, can easily be absorbed into the total costs of mineral extraction.

Salinity Saline soils are those with an excessive amount of dissolved salts so that crop production is reduced or prevented. In mild cases the effect may be only a slight reduction of yield. In more serious cases the choice of crops may be limited to those which can stand some degree of salinity. In the worst cases nothing is able to grow unless remedial action is

Plate 13.16 Buffel grass *Cenchrus ciliaris* seeded in pits on scald areas near Alice Springs. A tine pitter is used with an eccentric wheel for intermittent lifting. The grass is sown at the same time as pitting (Conservation Commission of the Northern Territory of Australia)

taken. Different chemical salts cause different kinds of salinity, and require different treatments. Soils which are low in total salt content but have a high proportion of sodium salts are known as *sodic* or *black alkali* or *solonetz*; soils with low sodium salts but high total salt content are called *saline* or *white alkali*; soils with both high sodium content and high total salt content are called *saline-alkali* or *solonchack*, or *sodic-alkali*.

The two treatment methods are leaching, that is flushing out the salts by the application and draining away of relatively clean water, and the application of chemicals such as calcium carbonate (lime or chalk), calcium sulphate (gypsum), or sulphur. Details are given in *Further reading*.

13.5 Control of streambank erosion

In terms of the area affected, streambank erosion is a small problem, but its significance increases when other factors are considered. Land subject to surface erosion is only partly damaged, but land lost by bank erosion is completely and irretrievably lost, and what is more, the bottom lands of valley floors are nearly always valuable and highly productive. Bank erosion may also threaten roads and bridges with much more serious economic consequences than the value of the lost land. Considered as a source of sediment polluting the stream, bank erosion is important because all the soil goes directly into the stream with no possibility of any of it being trapped or filtered out.

13.5.1 Choice of control methods

On large rivers the main damaging action is the scour of the river flow, undermining the banks and causing their collapse. The regulation and control of extreme floods is outside the scope of this book, and our concern is how to minimize the damage done by smaller more frequent floods. One of the factors to be considered is the variation in the level of the river or stream. A river whose flood height is several metres above the normal level of flow is going to require more extensive bank protection than one whose flood rise is small.

Although most damage to the banks results from scour by the stream flow, another cause is surface runoff over the edges of the bank, and this aspect is more important in the case of smaller streams. Particularly vulnerable points are where man-made concentrations of water are discharged into the stream such as from drains or from mechanical protection works.

The risk of damage is increased when cultivation takes place right up to the edge of the stream or watercourse. If the natural vegetation is left undisturbed along the edge, this both slows down the surface runoff on its way into the stream, and also the stream in flood has its edges flowing over ground protected by vegetation, not over ploughed land. In Zimbabwe there is legislation which prohibits all cultivation or disturbance of the natural vegetation within 30 metres of the flood level of any stream or river.

Another example of streambank damage being affected by agricultural use is the cultivation of vegetable gardens in the beds of seasonally dry rivers (plate 13.17). The apparently dry sand deposits in the river bed often contain plenty of water not far below the surface. This allows the production of dry-season vegetables in small gardens temporarily cultivated in the bed of the stream. Shallow wells dug in the sand also provide water for domestic use and for cattle. The result is the disturbance or removal of the natural vegetation in the stream bed and on the banks, and much traffic up and down the banks, so that when the seasonal floods arrive the rate of stream bank erosion is increased. There is no easy

Plate 13.17 Vegetable gardens can make use of the water below the surface of the dry stream bed, but the next floods will carry away large amounts of soil

solution to this problem for to restrict the use of ground below flood level would be to eliminate a useful source of food production.

Attempts to prevent or reduce streambank erosion fall into two groups: those directed towards making the floods less damaging, and those directed towards making the banks more resistant, and these will each be discussed in turn.

13.5.2 Controlling the streamflow

The principle is either to deflect the water away from vulnerable points, or to slow it down so that it is less erosive. Slowing the water may result in the deposition of sediment load, and this can further help protect vulnerable places. To deflect the flow away from the banks, jetties or groynes are built from the bank, either at right angles or pointing downstream. Vertical timber piling is often used in deep water, or stone or concrete blocks. The use of gabions or wire baskets filled with rocks is described in section 12.5.2 on gully control, and these are also very effective for building groynes. The main advantage is that they have enough flexibility to readjust to localized scour at the base when a rigid structure might crack.

An alternative approach is to build retards or permeable spurs, which stick out into the stream like jetties, but slow down the stream rather than divert it. Timber is the usual material, although prefabricated metal frames have been used in America. To save the laborious and expensive driving of piles into the river bed to anchor the retard it is common practice to hold it against the force of the water by wire cables fastened from the bank to the outer end of the structure. Illustrations and diagrams of the methods used in America are shown in AYRES (1936) and BENNETT (1939).

A modern version of *hard engineering* streamflow control is described by LAFAYETTE and PAWELEK (1989) for a project in a National Forest in New Mexico, USA. A porous revetment was built from pre-drilled galvanized steel posts and galvanized V-mesh wire fence fabric, and sited on the inside of a bend where the outside of the bend was eroding the bank (figure 13.5). In the gap between the bank and the revetment, porous baffles were built across the old channel using the same materials, and a new channel was cut on the inside of the revetment.

The porous barriers in the old channel slow down the flow, so decreasing the bank erosion and depositing sediment between the baffles, helped by planted vegetation. The new channel was stabilized by log drop structures and planting both banks to willows. The results have been closely monitored and are sufficiently encouraging for the Forest Service to extend the scheme to other locations.

13.5.3 Protecting the bank

Instead of controlling the streamflow, or perhaps in addition to such measures, the approach may be to protect the bank against erosion. Vegetation is unlikely to be successful on its own because the fast-flowing

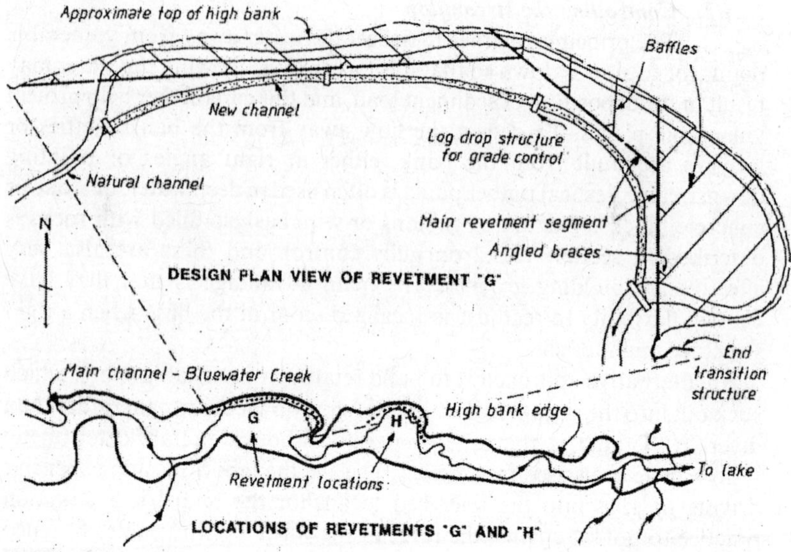

Figure 13.5 Managing river bank erosion through structures (from LAFAYETTE and PAWELEK 1989)

turbulent flow of a stream is much more erosive than the shallow smooth flow which can be kept to non-scouring velocities in a grassed waterway. However, when unstable banks are shaped to a gentler slope they require vegetation to protect them from rain erosion and surface runoff, and some protection against scour by the stream may also be obtained. The requirements of such vegetation are rapid growth and a dense shallow rooting system. Willows (*Salix*) are very suitable and can be used in several ways. Complete rooted bushes may be planted, or cuttings will strike easily in moist soil. A 60-year old technique described by AYRES (1936) is to bury willow poles in shallow trenches extending from just below low-water level and extending up the slope to the top of the bank. With the butts in water the poles develop a spreading root system, and send up shoots at every joint. The trenches are spaced about a metre apart, so a uniform cover soon develops.

Vegetation alone may not be sufficient protection against the scour of the stream, in which case mechanical protection is also required. One of the simpler methods is mats woven from flexible tree or bush cuttings, and willows are commonly used for this purpose. The banks are shaped to a smooth slope of 1 in 2 or flatter, and the mats are secured on the banks by wiring them down onto stakes driven through the mats. The mats are continued down below the water line, and the underwater parts are weighted down by large rocks or concrete slabs.

A combination of hard structure and growing vegetation may be useful where the vegetation needs help while becoming established. Two variations are described by KOHNKE and BOLLER (1989). The *live cribwall* (figure 13.6A) consists of box cribs, made from rough-cut untreated timber, resting on a rock foundation. The crib is filled with rock, soil, and cuttings of a range of bushes, shrubs and trees which flourish on river banks. The *live soft gabions* (figure 13.6B) are basket-like structures with successive layers of soil and planting material held in place with sheets of geotextile for strength, and a filter fabric to stop the soil being washed out.

Other materials for permanent protection works on sloping banks are masonry and concrete, both effective but expensive. The disadvantage of concrete, its lack of flexibility, can be overcome by using concrete panels joined together by flexible couplings. Dry walling built of selected undressed stones is called *rip-rap* in America. A cement grout can be brushed in if required. Gabions have been mentioned for gully control (section 12.5.2) and for jetties (section 13.5.2), and they can also be used to build vertical or stepped walls. For protecting sloping banks a variation is used called the *Reno mattress*. This is the same kind of wire mesh basket filled *in situ* with stones and rock, but is wider and thinner. Each unit is 6 m long by 2 m wide, with a choice of 170, 230 or 290 mm depth. Each unit is divided internally into 10 separate compartments each 0.6 m long, and this prevents the stone slipping down to the lower end (plate 13.18).

Yet another variation is the *sack gabion* which can be used on steeply sloping banks requiring protection above and below the water level. The empty gabions, shaped like huge sausages, are hung over the edge of the bank for ease of filling, and when full are lifted by crane and lowered into position on the bank, extending below the water level (plate 13.19).

References

ABEL, N. O. J. and P. M. BLAKIE 1989 Land degradation, stocking rates, and conservation policies in the communal rangelands of Botswana and Zimbabwe, *Land Degradation and Rehabilitation*, 1, 101–123

AYRES, Q. C. 1936 *Soil Erosion and Its Control*, McGraw-Hill, New York and London

BEHNKE, R. H. and I. SCOONES 1991 Rethinking Range Ecology: Implications for Rangeland Management in Africa, *Commonwealth Secretariat*, London, also published as Dryland Networks Programme, *Issues Paper 33*, 1992, IIED, London

BENNER, R. E. 1985 Urban sediment and stormwater control: The

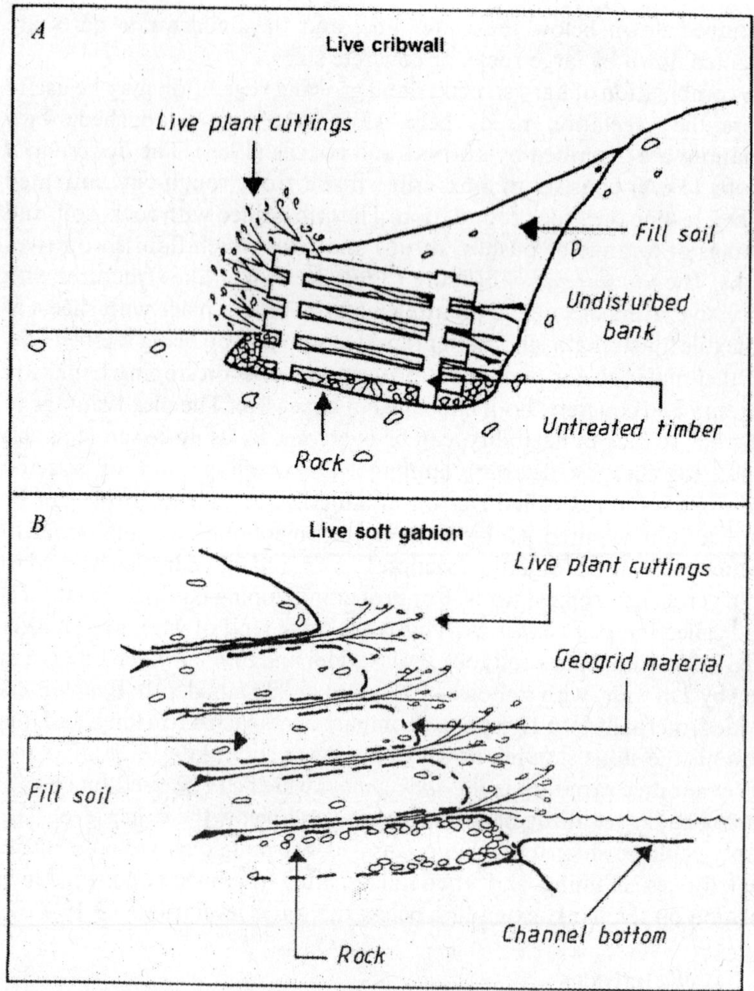

Figure 13.6 Bioengineering control of streambank érosion
A Live cribwall
B Live soft gabions (from KOHNKE and BOLLER 1989)

Plate 13.18 The 'Reno mattress' is a wide and shallow form of gabion used to give protection to sloping banks (RIVER and SEA GABIONS)

Plate 13.19 Sack gabions are used to protect banks in deep water. They are filled in the vertical position while hanging down the bank then lowered into position by crane (RIVER and SEA GABIONS)

Maryland experience, *Journal of Soil and Water Conservation*, 40, 1

BENNETT, H. H. 1939 *Soil Conservation*, McGraw-Hill, New York

BERG, W. A., C. L. DEWALD and P. L. SIMS 1992 Grass and forb establishment on highly erodible marginal farmland in the southern plains of the USA, chapter 34, in *Soil Conservation for Survival*, edited by H. Hurni and K. Tato, Soil and Water Conservation Society, Ankeny, Iowa

BIAMAH, E. K. 1992 Production oriented conservation strategy for developing sustainable projects in arid and semi-arid areas of Kenya, in *People Protecting their Land*, edited by P. G. Hoskins and B. M. Murphy, Department of Conservation and Land Management, Sydney, New South Wales

BRANDT, C. J. 1987 The effect of different types of forest management on the transformation of rainfall energy by the canopy in relation to soil erosion, *Forest hydrology and watershed management*, IAHS 167, 213–222

BRYANT, F. C., B. E. DAHL, R. D. PETTIT and C. M. BRITTON 1989 Does short-duration grazing work in arid and semi-arid regions? *Journal of Soil and Water Conservation*, 44, 4, 290–296

CHADHOKAR, P. A. 1992 Area Closure for Soil Conservation in Ethiopia: Potential and Dangers, chapter 26, in *Soil Conservation for Survival*, edited by H. Hurni and K. Tato, Soil and Water Conservation Society, Ankeny, Iowa

CHAPMAN, G. 1948 Size of raindrops and their striking force on the soil surface in a Red Pine Plantation, *Transactions of American Geophysical Union*, 29, 664

CHRISTIANSSON, C., I. S. KIKULA and W. OSTBERG 1991 Man-land interrelations in semi-arid Tanzania: A multidisciplinary research programme, *Ambio*, 20, 8, 357–361

COPPIN, N. J. and I. G. RICHARDS 1990 *The Use of Vegetation in Civil Engineering*, Construction Industry Research and Information Association (CIRIA), Book B10, Butterworths, London

CUSKELLY, S. L. 1969 Erosion-control problems and practices on National Forest Lands, *Transactions of the American Society of Agricultural Engineers*, 12, 1, 69–70, 85

DIXON, R. M. 1987 Imprintation: a process for reversing desertification, in Erosion Control – you're gambling without it, *Proceedings of 28th Annual Conference of IECA*, Steamboat Springs, Colorado, 297–306

FAO 1981 Arid Zone Hydrology, *Irrigation and Drainage Paper 37*, FAO, Rome

FAO 1987 Soil and water conservation in semi-arid areas, *Soils*

Bulletin 57, FAO, Rome

FAO 1989 Road design and construction in sensitive watersheds, *Conservation Guide 13, Watershed management field manual 13/5*, FAO, Rome

FERGUSON, B. K. 1990 Urban stormwater infiltration: purposes, implementation, results, *Journal of Soil and Water Conservation*, 45, 6, 605–609

FERGUSON, B. K. 1991 Taking advantage of stormwater control basins in urban landscapes, *Journal of Soil and Water Conservation*, 46, 2, 100–103

FIFIELD, J. S., L. K. MALNOR and L. E. DEZMAN 1989 Effectiveness of erosion control products on steep slopes to control sediment and to establish dryland grasses, in *Erosion knows no boundaries, Proceedings of 20th Conference of IECA*, 47–55, Steamboat Springs, Colorado

FOWLER, M. 1981 Overgrazing in Swaziland? A review of the technical efficiency of the Swaziland herd, *Pastoral Network Paper 12d*, ODI, London

FULLEN, M. A., A. M. TYE, D. A. PRITCHARD and H. A. REYNOLDS 1993 Effects of 'AGRI-SC' soil conditioner on the erodibility of loamy sand soils in East Shropshire, UK, *Soil Use and Management*, 9, 1, 21–24

GARDNER, R. B. 1979 Some environmental and economic effects of alternative forest road designs, *Transactions of the American Society of Agricultural Engineers*, 22, 1, 63–68

GICHANGI, E. M., R. J. JONES, D. M. NJARUI, J. R. SIMPSON, J. M. N. MUTUTHO and S. K. KITHEKA 1992 Pitting practices for rehabilitating eroded grazing land in the semi-arid tropics of eastern Kenya, chapter 25, in *Soil Conservation for Survival*, 313–327, edited by H. Hurni and K. Tato, Soil and Water Conservation Society, Ankeny, Iowa

GOLDMAN, S. J., K. JACKSON and T. A. BURSZTYNSKY 1986 *Erosion and Sediment Control Handbook*, 6.20–6.29, McGraw-Hill, New York

HARCHARIK, D. A. and S. H. KUNKLE 1978 Forest plantation for rehabilitating eroded lands, chapter 6, in *Special Readings in Conservation, Conservation Guide 4*, FAO, Rome

HARDIN, G. 1968 The Tragedy of the Commons, *Science*, 162, 1243–48

HOWELL, J., J. E. CLARK, C. LAWRANCE and I. SUNWAR 1991 Vegetation structures for stabilising highway slopes, *Department of Roads*, Kathmandu, Nepal

HUDSON, N. W. 1992 *Land Husbandry*, Batsford, London

HUGGARD, E. R. 1958 *Forest Engineering Handbook*, Heffer, Cambridge

HUNTER, J. P. and L. C. WEAVER 1993 Developing sustainable grazing associations in Lesotho, in *Working With Farmers for Better Land Husbandry*, edited by N. W. Hudson and R. J. Cheatle, Intermediate Technology Publications, London

JONES, R. M. 1966 Scald reclamation Studies in the Hay District, Part I Natural Reclamation of Scalds, Part II Reclamation by Ploughing, Part III Reclamation by Ponding Banks, Part IV Scald soils: Their Properties and Changes with Reclamation, *Journal of the Soil Conservation Service of New South Wales*, Part I 22, 3, 147–158; Part II 22, 4, 213–230; Part III 23, 57–64 (1967); Part IV 25, 1, 104–120 (1969)

KOHNKE, R. E. and A. K. BOLLER 1989 Soil bioengineering for streambank protection, *Journal of Soil and Water Conservation*, 44, 4, 286–287

LAFAYETTE, R. A. and D. W. PAWELEK 1989 New revetment design controls streambank erosion, in *Proceedings of 20th Conference of IECA*, Steamboat Springs, Colorado

LAWRANCE, C. J., P. G. CARTER and M. J. MAPPLEBECK 1991 Cylinder retaining walls, *Highways and Transportation*, June 1991, 30–34

MEGAHAN, W. F. 1977 Reducing Erosional Impacts of Roads in Guidelines of watershed management, *Conservation Guide 1*, FAO, Rome

MNDEME, K. C. H. 1992 Combating soil erosion in Tanzania: the 'HADO' experience, chapter 34, in *Soil Conservation for Survival*, edited by K. Tato and H. Hurni, Soil and Water Conservation Society, Ankeny, Iowa

MOSELEY, M. P. 1982 The effect of a New Zealand beech forest canopy on the kinetic energy of water drops and on surface erosion, *Earth Surface Processes and Landforms*, 7, 103–107

NEPAL SPECIAL PUBLIC WORKS PROGRAMME 1992 Environmental protection measures for hill irrigation schemes in Nepal, *SPWP Manual 1*, Department of Irrigation, Government of Nepal

PACEY, A. with A. CULLIS 1986 *Rainwater Harvesting – the collection of rainfall and runoff in rural areas*, Intermediate Technology Publications, London

PEREIRA, H. C., P. H. HOSEGOOD and D. B. THOMAS 1961 The productivity of tropical semi-arid thorn-scrub country under intensive management, *Empire Journal of Experimental Agriculture*, 29, 115, 269–286

PIEPER, R. D. and R. K. HEITSCHMIDT 1988 Is short-duration stocking the answer? *Journal of Soil and Water Conservation*, 43, 2, 133–137

PRATT, M. 1983 Land imprinting for lasting impressions, *Agricultural Engineering*, 64, 9, 8–10

QUILTY, J. A., J. S. HUNT and R. W. HICKS 1979 Urban Erosion and Sediment Control *Technical Handbook 2*, Soil Conservation Service of New South Wales

RICKSON, R. J. 1990 The Role of Simulated Vegetation in Soil Erosion Control, chapter 8, in *Vegetation and Erosion*, edited by J. B. Thornes, Wiley, Chichester, Sussex

SANDFORD, S. 1983 *Management of pastoral development in the Third World*, Wiley, Chichester, Sussex

SAVORY, A. 1988 *Holistic Resource Management*, Island Press, Washington DC

SCHIECHTL, H. M. 1985 Vegetative and soil treatment measures, *Watershed management field manual, Conservation Guide 13/1*, FAO, Rome

SHAINBERG, I., D. WARRINGTON and P. RENGASAMY 1990 Effect of soil conditioner and gypsum application on rain infiltration and erosion, *Soil Science Society of America Journal*, 54, 1104–1106

SHENG, T. C. 1978 Hydro-seeding: Procedures, Examples, and Prospects in Jamaica, chapter 5, in *Special Readings in Conservation, Conservation Guide 4*, FAO, Rome

SUWARDJO, H., A. DARIAH and A. BARUS 1991 Rehabilitation of degraded land in Indonesia, in *Development of Conservation Farming on Hillslopes*, edited by W. C. Moldenhauer, N. W. Hudson, T. C. Sheng, and San-Wei Lee, Soil and Water Conservation Society, Ankeny, Iowa

TAI, B. 1992 Erosion control on urban development sites: Compliance with best practice in the Sydney Region (NSW), in *People Protecting their Land*, edited by P. G. Hoskins and B. M. Murphy, Department of Conservation and Land Management, Sydney, New South Wales

TAYLOR, C. A. 1989 Short-duration grazing: Experiences from the Edwards Plateau region of Texas, *Journal of Soil and Water Conservation*, 44, 4, 297–302

UNDP 1984 Project Completion Report Eastern Senegal Livestock Development Project, Western Africa, *Projects Department (WAPAC)*, World Bank, Washington DC

US ENVIRONMENTAL PROTECTION AGENCY 1973 Comparative costs of erosion and sediment control construction activities, *EPA-430/9-73-016*, Washington DC

WALLACE, A. and G. A. WALLACE 1986 Effects of very low rates of synthetic soil conditioners on soils, *Soil Science*, 141, 324–327

14

Policies and implementation

14.1 Changes in policy and strategy

14.1.1 Policy changes

In the discussion on land use planning we suggested that the ideal pattern of development is to establish a long-term national or regional policy, which leads to medium-term strategies, and short-term tactical operations. In the real world a curious situation is arising in that the adoption of the new land husbandry approach is growing so fast that the practitioners have moved ahead of the politicians and opinion formers, with the result that the implementation of strategies by Non-Government Organisations, aid agencies, and senior field officers, is helping to shape new policies.

There is always the problem that national policies ought to be long term, but are created by politicians usually working within a short time frame to the next election. Fortunately, the land husbandry approach allows planners at all levels from farmers to governments to appreciate the long-term issues.

The new approach was mentioned in section 9.2.1 in connection with land use planning. It is summed up in the slogan *From soil conservation to land husbandry*, and two reviews using that title list the main points as the need for:

- Strong political support (an excellent example from Kenya is quoted in LUNGDREN 1993, BOX 14).
- Adequate resources to implement policies including finance and trained manpower.
- Adequate institutional strength from ministerial level to self-help farm groups, discussed further in 14.2.2.

At implementation level the changes are:

- From thinking 'soil conservation' to thinking 'land husbandry', which includes increased productivity and water management.
- From earth-moving soil conservation to better farming.
- From quick-fix projects to cure erosion to long-term programmes to prevent it.
- From top-down plans to people-led development.

14.1.2 Transferring technology

When a new technology is introduced and found to be very successful there is a tendency for it to be used widely, including in conditions outside those for which it was designed. One example is the Universal Soil Loss Equation which was extremely successful, but widely misused out of context. Another example is the classical package for controlling erosion and runoff by engineered channels (section 10.4.2 and figure 10.9). The philosophy and the techniques developed in the United States were too often transferred to the rest of the world in the 1930s and 40s without sufficient local testing, and it has taken us nearly 50 years to realize how much of that expertise is not suitable for developing countries. The reason is apparent if we compare the conditions in which soil conservation technology developed in the United States with conditions in developing countries.

In the United States of America the conditions are:

- They have, on the whole, good soils and a favourable climate, at least compared with most developing countries.
- Pressure on the land is low, so it is possible to concentrate production on high-potential areas and limit the use of marginal land.
- There is strong and sustained government support for the agricultural industry, which results in high-quality services for extension, research, and conservation.
- The farmers are on the whole well-informed, operating large mechanized farms with high capital investment, a sound financial base, and ready access to credit.
- They can expect reliable prices and market outlets for their produce, often maintained by government intervention.

Not one of these conditions applies to resource-poor, small-scale farmers in developing countries. The farmers' objectives are also very different. In the developed world the farmer is usually looking for maximum yield either per unit of land or per unit of money invested. The small-scale or subsistence farmer is more likely to be looking for security of food supply, a good return from the labour input, the reduction of drudgery, and a move to the cash economy. So it is not surprising that new approaches and new technologies are required in developing countries. The basic principles of subjects such as soil physics and soil chemistry, or plant selection and breeding, are universal but applying them is not.

14.1.3 Past failures

In the last 20 or 30 years huge sums of money have been spent on programmes of agricultural development either specifically directed to solving the problems of land degradation or including that as one component. The results have been disappointing, and national governments, national and international agencies, and NGOs have all become concerned. Several studies have been commissioned to determine the reasons for poor past performance in order to improve future programmes; examples are studies by the WORLD BANK (1984, 1985, 1986), USAID (1985), IFAD (1986), and FAO (1991).

A recent review identified nine common shortcomings (PRIOR 1992).

- A lack of beneficiary involvement.
- Failure to recognize the importance of indigenous farming systems.
- High recurrent cost, high maintenance requirement, and complex technologies.
- Failure to recognize the low absorptive capacity of local governments and communities.
- Failure to recognize the importance of land tenure issues.
- Failure to recognize the importance of livestock in the system.
- Absence of sustainability and social equity goals.
- Poor understanding of the primary causes of land degradation.
- Over-emphasis on the project approach (in which a project design is established at the outset) rather than the process approach in which the project develops as it learns (KORTEN 1980).

It is probably too optimistic to suggest that all of these lessons have been completely absorbed by the entire aid and development industry, because some of the institutions involved are rather like the proverbial oil supertanker which has so much momentum that there is a long delay between taking a decision on the bridge to change course and the effect becoming noticeable.

14.1.4 The way forward

Awareness of the low success rate has led to a number of changes in the approach to soil conservation projects.

- Projects dealing only or primarily with soil conservation have been largely abandoned in favour of broader projects where soil conservation is just one component in a programme for agricultural improvement.

- Programmes mainly dependent on mechanical soil conservation works have been virtually abandoned in favour of more emphasis on reducing erosion through improved farming methods.

- Programmes to improve agricultural production through large-scale mechanization have been abandoned as unworkable, and only survive in a few countries under socialist government.

- Large multi-sector programmes which need a separate administration unit are seldom started nowadays. They tend to undermine existing line departments and also leave a large gap at the end of the project. The preferred alternative is to work through existing departments in the public sector and local institutions in the private sector.

The requirements for an effective soil conservation policy for Europe were defined by MORGAN and RICKSON (1990) and with slight modification are equally relevant to policy and implementation elsewhere.

- There must be a high level of political pressure.
- There must be an available technology which has been shown to work.
- There must be a clear financial incentive. In the case of a policy this must accrue to government, in the case of projects it must accrue to the participating farmers.
- The policy must not be an end in itself, but part of an environment protection programme, or in the case of projects part of an agricultural development programme.
- There must be adequate local organizations to take responsibility for implementation.

To these broad generalizations we might add some details. It is generally accepted that it is desirable that there should be as little external intervention as possible, making the maximum use of existing social structure and institutions and existing farming practices, and minimizing the introduction of unfamiliar concepts or technologies. The use of subsidies or incentives should also be avoided if possible, because the intention is that the programme should be intrinsically attractive to the farmers and not require that their cooperation must be bought. The old style of blue-print projects with every detail worked out in advance is now giving way to more flexibility so that programmes or projects can be modified during implementation to take advantage of experience gained and lessons learned.

To review all the examples of programmes following this new approach would require a large book, but it might be appropriate to mention briefly two examples. The first comes from China: 'In a successful FAO project on the loess plateau of China, it was found necessary to introduce new systems of animal production, new horticultural crops, food processing, and better marketing facilities to the area. This allowed farmers to terrace steep hillsides and plant them to fruit trees and fodder plants rather than to grow wheat as they had done in the past. Through these and other measures, the value of agricultural production increased approximately fourfold in the pilot area, and erosion was reduced to an acceptable rate. Similar activities are now being expanded over a much wider area as a result of this pilot demonstration' (SANDERS 1992a).

The second example comes from Brazil, sponsored by the FAO/ Investment Center/World Bank Cooperative Programme in the States of Parana and Santa Catarina. The problem to be addressed was destructive land use practices adopted by settlers when they first deforested large areas, mostly in the 1950s or 60s. The purpose of the project was to encourage change to non-destructive sustainable systems of production by substituting land management practices which (a) protect the soil from surface sealing by rain splash, and (b) maintain an open structure for infiltration of rain water, accompanied by technical strategies including (c) mechanical or vegetative barriers to dispose safely of runoff from agricultural land or from roads after intense rain storms.

The specific problems facing the farmers were washouts of newly-seeded crops, flooding of low-lying areas used for livestock because of easy access to water, bacterial contamination of water supplies, and declining farm productivity and financial return. Key project components are extension, research on promising farm practices, road reshaping, and a soil conservation fund. This fund provides groups of farmers with part of the cost of upgrading land management, reforesting areas needing long-term protection, or to install cross-slope barriers. The basic unit for project implementation is the micro-catchment, typically of 3 000 to 5 000 ha with between 80 and 200 farmers. Central to each project is a joint agreement by all land users in the micro-catchment. An intensive extension programme is maintained for two years to form small neighbour groups, work out individual land use plans for each farmer, and consolidate the group plans through a committee. After this period of intensive support, there is a reduced level of support until the micro-catchment becomes independent and self-sufficient. The Parana project has now been operational for three years, the Santa Catarina project for one year, and a recent evaluation mission reported 'Project achievements are spectacular', and the programme is expanding rapidly into other states in Brazil, and to other countries in Latin America (FAO 1993).

14.2 Political aspects

14.2.1 Political commitment

Most national governments pay lip service to the idea of the conservation of natural resources, but in practice soil conservation is not a vote-winning issue with the electorate. The officially declared government policies are usually not translated into action because resources are allocated to more visible programmes which promise more immediate benefits. Long-term care of the soil cannot compete with the pressing need for schools, hospitals, or roads. The new approaches to policy planning could change this situation. The old-fashioned style of top-down conservation with a strong element of mechanical structures was always unpopular with the general public, but this need not apply to the new approach of land husbandry programmes which are attractive to the farmer. When the government can say: 'Look how we helped you to improve your living standards', land husbandry programmes will be vote winners.

A difficult question which can produce several conflicting answers is: 'Who is morally responsible for the care of natural resources?' One view is that this is a social responsibility of the state on the grounds that caring for the nation's land should be like the obligation to improve education and health care. But the administrative structure of governments which divides responsibility into ministries and departments is not appropriate for looking after natural resources as that requires a multi-sectoral and multi-disciplinary approach. Some countries have attempted to solve the question by creating a special body to bridge the gap between ministries, such as the Natural Resources Board which has functioned in Zimbabwe for 50 years, or the more recent example from Kenya of the Permanent Presidential Commission on Soil Conservation and Aforestation. Most governments have several ministries with an interest in natural resources and find difficulty in putting together collaborative programmes.

Another approach is to establish a sense of moral responsibility throughout the whole population, developed around the concept that it is morally wrong to maltreat the land in the same way that it is unacceptable to maltreat children. The conservation movement in the United States was based on the moral obligation of farmers to look after their land on behalf of future generations but it is hard to apply this in the face of economic pressures to increase production at the expense of soil degradation, and also the non-farm part of the population is less interested. The present Landcare movement in Australia aims to involve the whole of each rural community and this approach is showing considerable success but there is still the problem of involving the urban population.

The third approach is that those who most stand to gain from reducing land degradation must pay. This is relatively easy to apply to on-site damage but more difficult in the case of off-site problems. The sediment which is coming from mountains as a result of poor farming practices there can increase the cost of downstream irrigation, power generation, and water supply. It is unlikely that the hill farmer will be able or willing to pay for remedial measures, so should the bill be passed to the irrigation farmers? or the consumers of electricity or water? or the country as a whole through taxes? There are no clear-cut answers but a consensus on the role of the state is emerging. At the 1989 ISCO conference in Ethiopia a keynote address said: 'Although governments must take the overall responsibility for conservation, their main role should be to promote the participation of rural people in finding and applying the solutions. Policies and strategies must be orientated in this direction,' (SANDERS 1992b). This was further developed at the 1992 ISCO conference in Sydney where there was a consensus that: 'Control of land degradation must receive higher levels of attention and commitment by all governments if future populations are to be fed and sustained.' Another conclusion was that: 'The role of government has to change from being an organisation which "does conservation" to a body which encourages land users to protect their own land and creates conditions under which this will happen' (ISCO 1992).

A recent review of the present situation on soil and water management policies in more than ten countries in east and central Africa showed a strong correlation between political stability and having an effective programme for soil and water management. In some countries this aspect of government activity is in disarray as a result of political changes or civil unrest or tribal conflict. Other countries are making good progress because their policies and strategies create the right environment in which action for improvement can take place (HUDSON and CHEATLE 1993).

14.2.2 National and regional policies

The key features of policies were set out by EL-ASHRY (1988). 'Long-term actions, such as policy reform and institutional adjustment, coupled with long-term funding commitments, are the key to resolving many of the world's problems of resource management and agricultural productivity. These policies include the promotion of smallholder agriculture through the provision of producer incentives, rural credit, reasonable food crop prices, adequate market and transport systems, strong agricultural research and extension programmes and the promotion of farming systems that do not degrade land and water resources. These far sighted goals, however, require continuity within national governments and international assistance agencies.' In addition to these general principles there may well be special situations which affect national policy, for example the necessity to balance the need for food security against the desirability of export earnings from commodity crops. There may also be future contingencies to include in long-term planning such as when the revenue from oil runs out as in Venezuela and Nigeria, or the situation in which a large migrant labour force returns home as foreseen in Lesotho when labour is no longer welcome in South Africa, or in a number of countries in the Middle East who presently export much of their labour force to the oil-rich states of the Persian Gulf.

There is also the situation where: 'Governments do not have the capabilities or the political will to develop policies from long lists of possible actions, and need specific advice on what policies might work under the conditions in their countries, what economic benefits will result, and what institutional adjustments will be needed to carry out successfully these policy recommendations' (SANDERS, 1992b). This situation can be eased by the provision by international development agencies of an outline of policy guidelines applicable to a region, and a good example of this is the FAO document *The conservation and rehabilitation of African lands – an international scheme* (FAO 1990). Another source of guidance for national politicians arises from the interchange of ideas and experience among professional practitioners, international training workshops, and international conferences which today result in a significant feeding back of international experience into the preparation of national or regional policies.

14.2.3 Institutional constraints

The fashionable strategy for rural development today is to minimize intervention, and to maximize the use of existing organisations and institutions, developing these where necessary. An example of this is the marked swing away from Project Management Units (PMUs), common in the 1950s and 60s, which were newly created units set up to manage a project because that route was thought to be more effective than operating the project through existing line departments. One reason for adopting this approach is that in developing countries the existing institutional framework may not be strong enough to handle large new development programmes. Establishing PMUs was thought to be one solution, the other, adopted in the 1960s and 70s by the World Bank and other international agencies, was to make available huge loans to develop national agricutural research programmes or extension programmes where these were the constraint on agricultural development projects.

Even when the national agencies and institutions are adequate, the new style people-led grass-roots development can still be held back by the weakness or absence of management at District or Village level. An interesting example comes from Lesotho where government agricultural development policy is decentralization through the delegation of responsibility from the national Ministry of Agriculture to the Districts, where all relevant departments take part in the development of District Development Plans. Below this there has been set up a network of village councils and committees for all aspects of farming activity such as the Village Development Committee (VDC), the Land Allocation Committee (LAC), and Farmer Cooperatives. This well-intentioned scheme had to overcome two problems. The first was the traditional feudal hierarchy of king or paramount chief down through principle chiefs, senior chiefs, gazetted chiefs, and village headmen, which continues in parallel with the relatively democratic system of elected government with ministries and district commissioners. Taking decisions within the village instead of having instructions handed down was an unfamiliar concept. The second factor was that there was no experience among the farmers of managing committees – conducting meetings, keeping minutes and records, and keeping accounts. Training courses in basic book-keeping had to be the first step towards setting up village cooperatives which have been very successful. This is an example where a little assistance and advice from outside can be very effective. Key factors in the successful projects reported in *Working with Farmers for Better Land Husbandry* are stimulating cooperative action at village level, strengthening existing local institutions, and encouraging the formation of local self-help groups (HUDSON and CHEATLE 1993).

14.2.4 Legislation

The question of using persuasion or coercion to ensure appropriate land use is a continuing debate among conservationists, and of particular interest at the moment when the USA, so long a leader in the field of soil conservation, is moving in one direction, while most of the rest of the world moves in the opposite direction. In developing countries the present strong surge towards people-participation and people-led projects means that coercive legislation is being replaced by enabling legislation to provide the right ambience as discussed in the last section. Meanwhile in the USA the focus on a moral obligation to care for land is being faced by a mixture of two components. The 'green ticket strategy' is a combination of moral obligation reinforced by financial incentives. The 'red ticket strategy' includes several degrees of coercion, ranging from low-key pressure, a kind of moral blackmail to make people feel bad if they do not conform, through financial pressure applied by withholding subsidies unless conservation farming is practised, to the outright enforcement of legislation dictating land use. The key legislation was the Conservation Title of the Food Security Act of 1985 which for the first time in United States history included an element of coercion. Another new introduction was the principle of targeting, that is concentrating the conservation effort on defined problems, such as particularly erodible land, as opposed to the previous approach that assistance from the conservation department would be equally available to all. The effectiveness of this programme has been the subject of continuous debate since it was introduced, to the point where the Soil and Water Conservation Society recently devoted an entire edition of its journal to twelve papers discussing the subject (JSWC 48, 4, 1993).

A number of examples can be quoted to support or to undermine the case for legislation. From personal experience in what was then Southern Rhodesia, now Zimbabwe, as conservation officers we could theoretically call upon sweeping legislation to restrain overgrazing or exploitative land use, but we never needed to do so because public opinion and peer pressure was a much more effective regulator. Control through peer pressure is not confined to sophisticated societies; there are tribes in the Sahel where it would be an outrage to cut down an *Acacia albida* because it is so useful and respected.

There are also cases where authority enforced a particular practice without regard for whether or not it was acceptable, but in retrospect it has turned out to be justified. An example is the grass strips established by Royal decree in the Kingdom of Swaziland on all arable land in the early 1950s. They had been criticized because they were unpopular, and not maintained, and not always well laid out, but erosion would have been very much worse in Swaziland without them. The same applies to the contour bunds built during the days of colonial rule in many countries in Africa. With all their faults, they avoided the devastation of other countries where there was no control or restriction on exploitative land use such as Ethiopia.

The place of legislation in preserving forest is also debatable. The case for such legislation is that the state has a duty to care for the forestry sources and prevent them being damaged. But enforcing legislation which restricts the use of forests can turn the forest guardians into enemies of the people, and in many cases the well-meaning laws are unenforceable, which is always an undesirable situation. There are areas in northern India which are designated forest reserve, which theoretically allows no cutting, no grazing, and sometimes no entry, but on the ground there is hardly a tree to be seen, only eroded worn out grazing land (plate 14.1).

There are also examples of legislation intended to withhold land from settlement because it is ecologically unsuitable. This is particularly true of steep mountainous land in the humid tropics and there is today severe destruction occurring as the result of the spread of small-scale peasant farming onto land which cannot possibly support this use in the long term.

14.3 Social aspects

14.3.1 The effect of population pressure

In the past it has been generally held that a rapidly increasing population tends to cause overuse through an increasing demand being placed on a finite or decreasing resource. This was challenged by BOSERUP (1993) who argues that it is the increased demand which stimulates the necessity to produce more food and so encourages the development of new ideas or improved methods of production. Her thesis is carefully researched and well argued but it cannot overcome the problem that the comparisons have to be made between different situations, often in different countries, so one is never certain that like is being compared with like, nor is it possible to establish a cause and effect relationship. A most interesting recent study overcomes the first problem by concentrating on a particular area, the Machakos District in Kenya, where a really detailed database on the history of land use over 60 years provides the material for a detailed study of progressive change in the same area (TIFFEN *et al* 1993). This was a very serious study funded by the World Bank and involving a large team of specialist professionals over several years. The results are quite clear and show that without doubt there have been major changes in

Plate 14.1 This land in Himachal Pradesh, India, is nominally a protected forest reserve in which all tree cutting or cultivation is illegal

that soil erosion has been dramatically reduced, the yield in terms of both food production and cash income has increased both per hectare of land and per person on the land, while at the same time the population has increased fivefold over the 60 year period. But this study also raises questions – how much may the result have been influenced by the local conditions such as the fact that a main international railway line runs down one side of the district, there is also a busy international highway, and the proximity of the rapidly developing capital city of Nairobi influenced the outflow of labour and the inflow of capital for farm development? The second question, which has not been answered, is 'were these very important changes because of the increased population or in spite of it?' It is to be hoped that there will be many more serious studies of this topic.

An interesting question which is not resolved in either of these two studies is the interaction between population, livestock and land degradation. Throughout the developing countries of Africa much of the land degradation is the result of overgrazing and poor livestock management but there is a strong correlation between increasing human population and increasing livestock numbers. In most developing African countries the increase in numbers of cattle is equal to or greater than the increase in human population. In Machakos one of the main causes of the original degradation was overgrazing of communal lands, and this has been largely eliminated by land allocation and the change from free-range grazing to more intensive livestock management.

14.3.2 Labour and land husbandry

It is not clear whether there was ever any justification for the assumption that in developing countries labour is widely available at low cost. Certainly many early projects assumed this to be the case and the criteria for project success was the number of kilometres of ditches dug or bunds built. Quite recently *Food For Work* projects were based on bartering food for labour. Some were successful in reducing famine but few made constructive improvement to soil conservation.

It is quite certain that today farm labour is a valuable commodity in short supply, and this applies as much in developing countries as in Europe and North America. There are a number of possible reasons for this – a decrease in size of farm families; more children at school, and for longer; more opportunity for non-farm rural employment; and off-farm migration to towns or another country. An extreme example of the latter is the Yemen where a system of agriculture based on the labour-intensive construction and maintenance of terraces has virtually collapsed as a result of the mass migration of working males to the nearby Oil States (VOGEL 1987). Another example is Lesotho where much of the male labour force is drawn away from the poorly productive farm land by the higher wages in the mines and industries of the surrounding Republic of South Africa. It is not easy to see a solution to this because the standard of living of those left on the farm may be substantially higher as result of cash sent back by the migrants. Probably the process can only be slowed if there is more intensive cash-generating farming, or more rural industries processing farm produce.

How this affects the implementation of soil conservation practices is reviewed by STOCKING and ABEL (1992) who conclude that: 'All the evidence suggests that the availability of labour is a principle constraining factor in the acceptance or rejection of soil conservation. Labour-intensive techniques are only readily taken up and maintained on prosperous farms with regular cash-crop income. Elsewhere, soil conservation structures are fewer and in poor repair, even though farmers appreciate their value.' The return on labour may therefore be as important as the return on the investment of cash and this is discussed further in the next section.

14.3.3 Land tenure

Among the factors which lead to an over-stressing of land resources, we must include the cultural ethic that everyone has an automatic right to own land. This is not spelled out like 'the right to freedom of speech', but it lies at the heart of land management in most developing countries, and with some justification. Partly it is historical; in the past there was enough land for everyone to have some, and an increase in population just meant bringing more land into use. It also arises partly because in a dominantly agricultural economy there are no alternatives to working on the land.

Some forms of land tenure can lead to undesirable pressure on the land. Communal ownership can lead to mismanagement, particularly over-grazing by cattle, or the over-enthusiastic removal of firewood. Sometimes the existing system is difficult to change because there are vested interests at stake, such as the Chief or Headman having special rights. Where vested interests are not a problem it may be possible to move the land management towards a system which will bring greater rewards to the community as a whole. SHAXSON (1981) quotes such a case in central India. After many years of patient groundwork by extension workers, the village decided to rest and restore their main hill-grazing area, and operated a self-imposed and self-regulated programme of zero grazing and intensive planting of fuelwood trees.

The question of whether individual land ownership is essential for the implementation of conservation programmes through better land husbandry is producing conflicting reports. Social structures vary greatly and are usually complex so it is unlikely that any rationale will be generally applicable but some of the problems can be identified.

- Without tenure there is little incentive for the occupant to introduce long-term improvements.
- It can be a disadvantage when the land cannot be used as collateral for improvement loans.
- In some communities it is not possible for women to inherit tenure even though they may be the sole manager as a result of being widowed or the husband working elsewhere.
- In some communities the occupier may have tenure which is limited to crops during the growing season, with reversion to communal rights after the crop has been harvested. This inhibits the planting of trees and hedge plants on the boundaries of the arable land, and the use of agroforestry systems such as alley cropping on the crop land.

Plate 14.2 Working with the farmers is the key element of today's approach to soil conservation (WORLD NEIGHBORS)

- In some cases the requirement is not so much freehold land tenure as security of use or security of access which may be important for the management of grazing, or the planting of trees on communal land.

One important aspect of the change from soil conservation by specialists to inter-disciplinary land husbandry is the recognition that it requires a greater contribution than there has been in the past from social scientists, anthropologists, and economists. This is particularly true in trying to come to grips with tenurial rights and the role of women in the rural economy. In a recent discussion of the changing roles for rural sociologists TOBISSON (1993) points out that: 'A common conception in governments and international agencies is that rights to natural resources must be properly registered, and that economic progress will only materialize under the conditions of private property rights. The underlying notion is that traditional tenure systems lack security and that the land is unsuitable for long-term investments. However, anthropological observation suggests that traditional tenurial systems generally provide enough security for individuals and groups, while the farmer whose land rights are subject to modern law and tenure is often beset by insecurity.'

Tobisson also stresses the complexity of women's role and rights. 'Gender is a characteristic that helps us to analyse the internal dynamics of a community. An understanding of the linkages between gender, poverty, and the environment is necessary if we are to understand environmental processes and changes. There are direct linkages between women's and men's roles in the social and economic processes on one hand, and natural resource processes on the other. One example is land tenure, where men in most cases hold those rights and where women obtain usufructuary rights to family and kin resources through marriage. These rights may be invisible to policy makers, implying a risk that they may be eliminated in efforts to make land tenure regulations more formal and secure for individuals.'

14.4 Economic aspects

We cannot just assume that land degradation is 'a bad thing', and that something should be done about it without considering the realities of economics. Any programme intended to improve land husbandry will have costs, whether it is building terraces or stimulating self-help groups, and choices have to be made between alternative approaches. Making the best decisions requires being able to compare the effectiveness of alternatives, and for this we have to turn to economic analysis, assisted by specialists from other disciplines such as social science. Environmental economics is mushrooming today, and so is the literature, but much of it is economists talking to other economists so in this review we will make some probably invidious selections of papers which may help non-economists.

A constantly recurring theme is the difficulty of establishing the costs and the benefits. Programmes which involve the expenditure of funds whether national, bilateral, or international have to decide how to use the available funds to best advantage. But it is not easy to decide on relative priorities solely in economic terms, nor is it simple to convert all the benefits into cash values. It is fairly easy to calculate costs of soil conservation: more difficult to determine the benefits, even more difficult to quantify the benefits in cash terms. With some fairly subjective guesswork one can make an estimate and put cash values on increased production. With rather dubious calculations one can compute a value from not losing the soil. Secondary benefits like job creation and injecting money into the economy can be included, but there will always be at the end some intangible benefits which cannot be measured in cash values. There is no way of putting a cash value on the improved quality of life which comes from better health or a more balanced diet, or reduced drudgery among the rural population. A review of the basic features of economic analysis applied to soil degradation is given by BISHOP (1992) but TOBISSON (1993) argues that it is not appropriate to use neo-classical economics where the assumptions are about individual preferences and profit maximization. Peasant economies do not operate according to these economic laws. For example, risk minimization and family subsistence, rather than profit maximization, constitute a fundamental principle. Economic analysis of development should therefore be broader than the conventional cost-benefit analysis.

A contributor to the application of cost-benefit analysis (CBA) to land husbandry is Jan Bojö who developed ideas in Africa (1986), before joining the World Bank team studying this question. His review of 20 studies from all over the world of CBA applied to soil and water conservation confirms that identification of project costs is no problem but 'The identity of benefits occasionally poses problems', and that this should be addressed by a concerted research effort. He concludes that the actual application of CBA to conservation projects leaves many things to be desired, but that there are good examples of successful application, and that it could be a useful tool in the appraisal and evaluation of soil and water conservation projects (BOJÖ 1992).

A detailed on-farm economic analysis was carried out to compare the effectiveness of two different approaches to soil conservation implementation in the Trans-Nzoia District of western Kenya – the on-farm approach where soil conservation is voluntary with extension services and implementation on an individual basis, and the catchment approach with imposed heavy top-down soil conservation implemented on a cooperative basis throughout the whole catchment (EKBOM 1992). The result, showing that the catchment approach was more effective in this particular situation, is less important than the methodology which is an interesting move towards meeting Tobisson's point that the community perspective of effectiveness is what matters, rather than effectiveness from the perspective of the project.

This concept of assessing effectiveness from the viewpoint of the beneficiaries is continued in a study of terracing in the Machakos District of Kenya by TJERNSTROM (1992). There has been a long-term programme of soil and water conservation implemented by the Ministry of Agriculture in Kenya with strong support from the Swedish International Development Authority (SIDA), who have regularly evaluated their contribution to the programme. A much-studied question is the effect of terracing on crop yield, and Tjernstrom's study applied CBA first at the project level, using appropriate evaluation criteria and costs and benefits and showed a remarkably high internal rate of return (IRR) which he explains because the costs are very low in comparison with the benefits so in real money the return is not extremely great. A similar analysis at farm level, with particular attention to the problems faced by women farmers, showed that the rate of return in terms of cash per man-day was poor under a no-conservation regime, doubled under implementation of soil conservation at the present average level, and doubled again under full implementation of all the components of the recommended soil conservation package.

Two computer models developed in the USA allow a simplified form of CBA to compare the cost-effectiveness of alternative combinations of soil conservation practices. Both use the assumed reduction of soil loss as a measure of the benefits, calculated by applying the Universal Soil Loss Equation before and after implementation of each conservation practice.

'COSTS' is a short-term programme which costs 482 combinations of alternative rotations, tillage practices, and structural works such as terracing, and calculates the reduction in soil loss, then gives the farmer a display or print-out of the cost, the soil loss, and the cost per ton of soil saved (RAITT 1983).

'SOILEC' is a long-run physical and economic simulation model which like COSTS allows the farmer to compare the immediate cost-effectiveness of alternative management decisions, but also allows policy makers to assess the long-term effects over up to 50 years. In SOILEC the calculated soil loss is converted into the cash value of reduced yield and so gives an estimate of the financial return on costs. While the principle of these models may have general application they cannot be extrapolated to other situations because they depend firstly on the inputs from a huge file of data on cereal farming on local soils, and secondly on the accuracy of the estimates of soil loss and the resulting reduction of yield, both of which would need to be established before applying the model to different conditions (ELEVELD *et al* 1983).

References

BISHOP, J. 1992 *Economic Analysis of Soil degradation*, Gatekeeper Series, LEEC GK 93-01, IIED, London

BOJÖ, J. 1986 *An introduction to cost-benefit analysis of soil and water conservation projects*, Report 6, SADCC Soil and Water Conservation Coordination Unit, Maseru, Lesotho

BOJÖ, J. 1992 Cost-benefit analysis of soil and water conservation projects: A review of 20 empirical studies, chapter 18, in *Soil*

Conservation for Survival, edited by K. Tato and H. Hurni, Soil and Water Conservation Society, Ankeny, Iowa

BOSERUP, E. 1993 *The conditions of agricultural growth*, originally published in 1965, now reprinted by Earthscan Publications, London

EKBOM, A. 1992 *Economic impact assessment of implementation strategies for soil conservation*, Studies in Environmental Economics and Development, Gothenburg University, Sweden

EL-ASHRY, M. T. 1988 Foreword in *Conservation Farming on Steep Lands*, edited by W. C. Moldenhauer and N. W. Hudson, Soil and Water Conservation Society, Ankeny, Iowa

ELEVELD, B., G. V. JOHNSON and R. G. DUMSDAY 1983 SOILEC. Simulating the economics of soil conservation, *Journal of Soil and Water Conservation*, 38, 5, 387–389

FAO 1990 *The Conservation and Rehabilitation of African Lands: An International Scheme*, ARC/90/40, FAO, Rome

FAO 1991 A study of reasons for success or failure of soil conservation projects, *Soils Bulletin 64*, FAO, Rome

FAO 1993 *World Bank Land Management Projects in Parana and Santa Catarina, Brazil*, Working Paper: A synthesis of supervision mission observations, 6/93 CP-BRA52(WP), FAO, Rome

HUDSON, N. W. and R. J. CHEATLE (editors) 1993 *Working with Farmers for Better Land Husbandry*, Intermediate Technology Publications, London

IFAD 1986 *Soil and Water Conservation in Sub-Saharan Africa: Issues and Options*, Centre for Development Cooperation Services, Free University of Amsterdam

ISCO 7 1992 *Post-Conference Communiqué*, Department of Conservation and Land Management, Sydney, New South Wales

KORTEN, D. C. 1980 Community organisation and rural development: A learning process approach, *Public Administration Review*, Sept/Oct 1980

LUNDGREN, L. 1993 *From Soil Conservation to Land Husbandry*, Natural Resources Management Division, SIDA, Stockholm

MORGAN, R. P. C. and R. J. RICKSON 1990 Issues in Soil Erosion in Europe: the need for a Soil Conservation Policy, chapter 40, in *Soil Erosion on Agricultural Land*, edited by J. Boardman, I. D. L. Foster and J. A. Dearing, Wiley, Chichester, Sussex

PRIOR, J. C. 1992 Evaluating the experience of Third World soil and water conservation projects – nine common shortcomings, in *People Protecting their Land*, edited by P. G. Hoskins and B. M. Murphy,

Department of Conservation and Land Management, Sydney, New South Wales

RAITT, D. D. 1983 COSTS: Selecting cost-effective and conservation practices, *Journal of Soil and Water Conservation*, 38, 5, 384–386

SANDERS, D. W. 1992a Developing National and Regional Conservation Policies, chapter 1, in *Conservation policies for sustainable hillslope farming*, edited by S. Arsyad, I. Amien, T. C. Sheng and W. C. Moldenhauer, Soil and Water Conservation Society, Ankeny, Iowa

SANDERS, D. W. 1992b Soil Conservation: Strategies and Policies, chapter 3, in *Erosion, Conservation, and Small-scale Farming*, edited by H. Hurni and K. Tato, University of Berne, Switzerland

SHAXSON, T. F. 1981 Reconciling social and technical needs in conservation work on village farm lands, in *Soil Conservation: problems and prospects*, edited by R. P. C. Morgan, Wiley, Chichester, Sussex

SOIL AND WATER CONSERVATION SOCIETY 1989 Compliance and Conservation, special edition, *Journal of Soil and Water Conservation*, 44, 5

SOIL AND WATER CONSERVATION SOCIETY 1993 The next generation of US agricultural conservation policy, special edition, *Journal of Soil and Water Conservation*, 48, 4

STOCKING, M. and N. ABEL 1992 Labour Costs: A critical element in soil conservation, chapter 19, in *Soil Conservation for Survival*, edited by K. Tato and H. Hurni, Soil and Water Conservation Society, Ankeny, Iowa

TIFFEN, M., M. MORTIMORE and F. GICHUKI 1993 *More people, less erosion: environmental recovery in Kenya*, Wiley, Chichester, Sussex

TJERNSTROM 1992 Yields from terraced and non-terraced fields in the Machakos District of Kenya, chapter 23, in *Soil Conservation for Survival*, edited by K. Tato and H. Hurni, Soil and Water Conservation Society, Ankeny, Iowa

TOBISSON, E. 1993 Changing Roles for Rural Sociologists, in *Working with Farmers for Better Land Husbandry*, edited by N. W. Hudson and R. J. Cheatle, Intermediate Technology Publications, London

USAID 1985 *A soil and water conservation project in Somalia: seventeen years later*, Aid Project Impact Evaluation Report 62, USAID, Washington DC

VOGEL, H. 1987 Terrace farming in the Yemen Arab Republic. Traditional forms of soil and water conservation and their present degradation: a case study of the Manakhah Region, in *Soil Conservation and Productivity*, Vol 1, 585–607, edited by I. Pla Sentis, Soil Science Society of Venezuela, Maracay

WORLD BANK 1984 *Annual Review of Project Performance Results*, 10th Report 5248, Vol III

Appendix 1

Linear measure

1 kilometre	= 0·6214 mile	1 mile	=	1·609 kilometre
1 metre	= { 39·37 inches	1 yard	=	0·9144 metre
	3·2808 feet	1 foot	=	0·3048 metre
1 millimetre	= 0·03937 inch	1 inch	=	25·4 millimetres

Square measure

1 square kilometre	= 0·3861 square mile	=	247·1 acres
1 hectare	= 2·471 acre	=	107640 square feet
1 square metre	= 10·764 square feet	=	1·196 square yard
1 square centimetre	= 0·155 square inch		
1 square millimetre	= 0·00155 square inch		
1 square mile	= 2·5899 square kilometres		
1 acre	= 0·4047 hectare		
1 square yard	= 0·836 square metre		
1 square foot	= 0·0929 square metre		
1 square inch	= 645·2 square millimetres		

Cubic measure

1 cubic metre	= 35·314 cubic feet	=	1·308 cubic yard
1 cubic yard	= 0·7645 cubic metre		
1 cubic foot	= 0·02832 cubic metre	=	28·317 litres
1 cubic inch	= 16·38716 cubic centimetres		

Weight

1 metric ton	= 0·9842 ton (of 2240 pounds)	=	2204·6 pounds
	= 1·1023 ton (of 2000 pounds)		
1 kilogram	= 2·2046 pounds	=	35·274 ounces
1 ton (of 2240 pounds)	= 1·016 metric ton	=	1016 kilograms
1 pound	= 0·4536 kilogram	=	453·6 grams

Measures of Soil Loss

1 metric tonne/hectare	= 0·446 tons/acre (2000 lb)
	= 0·4 tons/acre (2240 lb)
1 kilogram hectare	= 0·895 lb acre

Appendix 2

Common and Botanical Names of some grasses and plants commonly used in soil conservation

A Grasses used for lining channels and waterways

Common name	Botanical name
Bahia grass	*Paspalum notatum*
Bermuda grass in America, or Star grass in Africa	*Cynodon dactylon*
Blue grama	*Bouteloua gracilis*
Buffalo grass in Australia, and sometimes in America, also called St Augustines grass in West Africa and Crab grass in Jamaica	*Stenotaphrum secundatum*
Buffalo grass in America can also mean Canary grass	*Buchloe dactyloides* *Phalaris canariensis*
Carpet grass	*Axonopus compressus*
Centipede grass	*Eremochloa ophiuroides*, or *Chrysopogon aciculatus*
Couch grass, or quick grass	*Cynodon* spp
Guinea grass	*Panicum maximum*
Kentucky bluegrass or Smooth-stalked meadow grass	*Poa pratensis*
Kikuyu grass	*Pennisetum clandestinum*
Pangola grass	*Digitaria decumbens*
Smooth brome grass	*Bromus inermis*
Swaziland finger grass	*Digitaria swazilandensis*
Tall fescue	*Festuca elatior* (or *Arundinacea)*

Weeping love grass	*Eragrostis curvula*
Wheatgrass	*Agropyron* spp

B Some plants used in gully control

Bluestem grasses	*Andropogon* spp
Calapagonium	*Calapagonium mucunoides*
Centrosema	*Centrosema pubescens*
Grama grasses	*Bouteloua* spp
Kudzu vine	*Pueraria thunbergiana*
Tropical Kudzu	*Pueraria phaseoloides*
Taiwan Kudzu	*Pueraria tonkinensis*
Lespedezah	*Lespedeza sericea* *Lespedeza juncea*
Reed canary grass	*Phalaris arundinacea*
Common reed	*Phragmytes* spp
Saltbush	*Atriplex* spp
Sand-bar willow	*Salix exigua*

C Grasses used for stabilizing banks of terraces

Bana grass	*Pennisetum purpureum* × *P. americanum*
Donkey grass (humid areas)	*Panicum trichlocladum*
Guatemala grass	*Tripsacum laxum*
Nandi Seteria	*Seteria anceps*
Napier grass	*Pennisetum purpureum*
Makarikari (dry areas)	*Panicum coloratum Makarikariensis*

D Leguminous cover crops (between orchard terraces)

Calapogonium mucunoides
Calapogonium caeruleum
Desmodium ovalifolium
Pueraria javanica

Abbreviations used in the text

ADP	Agricultural Development Plan
ARS	Agricultural Research Service
BLM	Bureau of Land Management (USA)
CAP	Common Agricultural Policy
CBA	Cost/Benefit Analysis
CONEX	Department of Conservation and Extension (Zimbabwe)
ERR	Economic rate of return
ESSC	European Society for Soil Conservation
FAO	Food and Agricultural Organization of the United Nations
FFW	Food for work
FSR	Farming Systems Research
GIS	Geographic Information System
IBSRAM	International Board for Soil Research and Management
ICRAF	International Centre for Research in Agriculture
ICRISAT	International Crops Research Institute for the Semi-Arid Tropics
IECA	International Erosion Control Association
IFAD	International Fund for Agricultural Development
IFIAS	International Federation of Institutes of Advanced Study
IIED	International Institute for Environment and Development
IITA	International Institute for Tropical Agriculture
ILCA	International Livestock Centre for Africa
IRD	Integrated Rural Development
ISCO	International Soil Conservation Organization
ISRIC	International Soil Reference and Information Centre
ISSS	International Soil Science Society
MAR	Mean annual rainfall
NASA	National American Space Authority

NGO	Non-government organization
ODA	Overseas Development Administration
PMU	Project Management Unit
PRA	Participatory Rural Appraisal
RCA	Resources Conservation Act
RRA	Rapid Rural Appraisal
RUSLE	Revised Universal Soil Loss Equation
SADC	Southern African Development Community
SCS	Soil Conservation Service
SIDA	Swedish International Development Authority
T&V	Training and Visits system
USAID	United States Agency for International Development
USDA	United States Department of Agriculture
USLE	Universal Soil Loss Equation
WEPP	Water Erosion Prediction Project

Further reading

Proceedings of conferences of the International Soil Conservation Organization (ISCO)

ISCO 1 Ghent, Belgium, March 1978, *Assessment of Erosion*, editors M. de Boodt and D. Gabriels, 1980, Wiley, Chichester, Sussex

ISCO 2 Silsoe, Bedford, England, July 1980, *Soil Conservation: Problems and Prospects*, editor R. P. C. Morgan, 1981, Wiley, Chichester, Sussex

ISCO 3 Honolulu, Hawaii, January 1983, *Soil Erosion and Conservation*, editors S. A. El-Swaify, W. C. Moldenhauer and A. Lo, 1985, Soil Conservation Society of America, Ankeny, Iowa

ISCO 4 Maracay, Venezuela, November 1985, *Soil Conservation and Productivity* (2 volumes), editor Ildefonso Pla Sentis, 1987, Soil Science Society of Venezuela

ISCO 5 Bangkok, Thailand, January 1988, *Land Conservation for Future Generations* (2 volumes), editor Sanarn Rimwanich, 1989, Dept. of Land Development, Ministry of Agriculture, Thailand

ISCO 6 Nairobi, Kenya, and Addis Ababa, Ethiopia, November 1989, Volume I *Soil Conservation for Survival*, editors Kebede Tato and H. Hurni, 1992, Soil and Water Conservation Society, Ankeny, Iowa, Volume II *Erosion, Conservation, and Small-scale Farming*, editors H. Hurni and K. Tato, 1992, University of Berne, Switzerland

ISCO 7 Sydney, Australia, 1992, *People Protecting Their Land* (2 volumes), 1992, Dept. of Conservation and Land Management, Sydney, New South Wales, Australia

ISCO 8 Held at New Delhi, India, December 1984

Proceedings of some international workshops

Workshop on Soil and Water Conservation on Steep Lands, San Juan, Puerto Rico, March 1987, *Conservation Farming on Steep Lands*, editors W. C. Moldenhauer and N. W. Hudson, 1988, World Association of Soil and Water Conservation, Ankeny, Iowa

International Conference on Steepland Agriculture in the Humid Tropics, Kuala Lumpur, Malaysia, August 1987, *Steepland Agriculture in the Humid Tropics*, editors T. H. Kay, A. M. Mokhtarrudin and A. B. Kahari, 1987, Malaysian Agricultural Research and Development Institute (MARDI) and Malaysian Society of Soil Science

Workshop on Soil Erosion on Agricultural Land, Coventry, January 1989, *Soil Erosion on Agricultural Land*, editors J. Boardman, I. D. L. Foster and J. A. Dearing, 1990, Wiley, Chichester, Sussex

Workshop on Conservation Farming on Hillslopes, Taichung, Taiwan, March 1989, *Development of Conservation Farming on Hillslopes*, editors W. C. Moldenhauer, N. W. Hudson, T. C. Sheng and San-Wei Lee, 1991, Soil and Water Conservation Society, Ankeny, Iowa

International Workshop on *Conservation Policies for Sustainable Hillslope Farming*, Solo, Indonesia, March 1991, editors S. Arsyad, I. Amien, T. C. Sheng and W. C. Moldenhauer, 1992, Soil and Water Conservation Society, Ankeny, Iowa

International Workshop on Soil and Water Management for Sustainable Smallholder Development, Arusha, Tanzania and Taita Hills, Kenya, June 1991, *Working with Farmers for Better Land Husbandry*, editors N. W. Hudson and R. J. Cheatle, 1992, Intermediate Technology Publications, London

International Symposium on *Water Erosion, Sedimentation, and Resource Conservation*, 1990, Dehradun, Central Soil and Water Conservation Research and Training Institute, Dehradun, India

FAO Soils Bulletins

32 A framework for land evaluation 1976

34 Assessing soil degradation 1977

52 Guidelines: land evaluation for rainfed agriculture 1983

54 Tillage systems for soil and water conservation 1984

57 Soil and water conservation for semi-arid areas 1987

60 Soil conservation for small farmers in the humid tropics 1989

64 Reasons for success and failure of conservation projects 1991

68 Field measurement of soil erosion and runoff 1994

FAO Conservation Guides

1 Guidelines for watershed management 1977
2 Hydrological techniques for upstream conservation 1976
3 Conservation in arid and semi-arid zones 1976
4 Special readings in conservation 1978
8 Management of upland watersheds 1983
13 Watershed Management Field Manual
 13/1 Vegetative and soil treatment measures 1985
 13/2 Gully control 1987
 13/3 Slope treatment measures and practices 1986
 13/6 Watershed survey and planning 1990

Some other useful background reading

General-interest publications

Land, Food, and People, 1984, Economic and Social Development Series 30, FAO, Rome

How good the earth, 1991, Land and Water Development Division, FAO, Rome

Protect and Produce, (2nd edition) 1992, Land and Water Development Division, FAO, Rome

Soil and Water Conservation in Sub-Saharan Africa, 1992, Centre for Development Cooperation Services, Free University of Amsterdam, for International Fund for Agricultural Development, Rome

Chapter 2 Erosion processes

Rill erosion; Processes and Significance, 1987, editor R. B. Bryan, Catena Supplement 8, Catena Verlag, Cremlingen, Germany

Processes and mechanics of erosion, 1986, R. P. C. Morgan, chapter 2, in Soil Erosion and Conservation, Longman, Harlow, Essex

Chapter 2 Loss of productivity

Soil erosion and crop productivity, 1985, edited by R. F. Follett and B. A. Stewart, American Society of Agronomy, Madison, Wisconsin

Proceedings of soil erosion and productivity workshop, 1990, edited by W. E. Larson, G. R. Foster, R. R. Allmaras and C. M. Smith, University of Minnesota, St Paul, Minnesota

Erosion and soil productivity: a review, 1984, M. A. Stocking, Consultant's Working Paper 1, AGLS, FAO, Rome

Effects of soil erosion on crop productivity, 1987, R. Lal, Critical Reviews in Plant Sciences, 5, 4, 303–367, CRC Press

Chapter 7 Models

Agriculture Nonpoint Source Pollution: Model Selection and Application, 1986, edited by A. Giorgini and F. Zingales, Elsevier, New York

Modelling and measuring catchment scale erosion and deposition, 1987, R. D. Barling, G. J. Burch, R. B. Grayson and A. K. Turner, in Steepland Agriculture in the Humid Tropics, Proceedings of 1987 Malaysia conference

Management, Conservation, and Erosion Data Base, 1988, T. M. Loran, J. A. Zinck and K. J. Beek, in Proceedings of ISCO 5, Thailand, Vol 1, 37–62

Erosion prediction models for large catchments, 1990, R. J. Garde and U. C. Kothyari, in Proceedings of 1990 Dehradun conference

Erosion processes and models, 1992, S. A. El-Swaify and J. H. Fownes, chapter 11, in Vol 1 of Proceedings of ISCO 6, Ethiopia

Chapter 7 Erosion and productivity

Erosion Impacts, 1985, Part II, chapters 21 to 28, in Proceedings of ISCO 3, Hawaii

Chapter 8 Rainfall simulators

Proceedings of the Rainfall Simulator Workshop, Tucson, Arizona, March 1979, ARM-W-10, USDA-SEA

Rainfall Simulation, Runoff, and Erosion, 1983, editor J. De Ploey, Catena Supplement 4, Catena Verlag, Cremlingen, Germany

Rainfall Simulators for Soil Conservation Research, 1988, L. D. Meyer, in Soil Erosion Research Methods, editor R. Lal, Soil and Water Conservation Society, Ankeny, Iowa

Chapter 9 Rapid rural appraisal

An introduction to Rapid Rural Appraisal for Agricultural Development, 1988, J. A. MacCraken, J. Pretty and G. R. Conway, IIED, London

Chapter 10 Manuals on soil conservation measures

Watershed development with special reference to soil and water conservation, 1979, N. Gil, FAO Soils Bulletin 40

Soil and water conservation in semi-arid areas, 1987, N. W. Hudson, FAO Soils Bulletin 57

Slope treatment measures and practices, 1988, FAO Conservation Guide 13/3

Vanishing land and water: Soil and Water conservation in dry lands, 1988, J. L. Chleq and H. Dupriez, Macmillan, London

Soil conservation for small farmers in the humid tropics, 1989, T. C. Sheng, FAO Soils Bulletin 60
Looking after our land, 1991, W. Critchley, Oxfam Publications

Chapter 11 Land husbandry

From Soil Conservation to Land Husbandry, 1993, L. Lundgren and G. Taylor, Natural Resources Management Division, SIDA, Stockholm
Conservation Farming, 1985, edited by P. E. V. Charman, Soil Conservation Service of New South Wales, Australia

Chapter 11 Conservation tillage

Tillage systems for soil and water conservation, 1984, P. W. Unger, FAO Soils Bulletin 54
Soil Tillage and Agricultural Sustainability, 1991, Proceedings of 12th Annual Conference of International Soil Tillage Research Organization (ISTRO), Ibadan, Nigeria
Conservation Tillage for Sustainable Crop Production Systems, collaborative programme between GTZ (Germany) and Dept. of Agricultural and Technical Services (AGRITEX), Ministry of Agriculture, Zimbabwe, Project Research Reports 1–7 (continuing), 1991–1993

Chapter 11 Agroforestry

Agroforestry for Soil Conservation, 1985, B. Lundgren and P. K. R. Nair, chapter 68, in Proceedings of ISCO 3, Hawaii
Agroforestry for Soil Conservation, 1989, A. Young, CAB International, Wallingford, Oxfordshire

Chapter 13 Grazing land

New directions in African range management policy, 1992, R. H. Behnke, Final Report Project FRD/7/105, Commonwealth Secretariat, London

Chapter 13 Roads and erosion

Control of erosion and sediment deposition from construction of highways and land development, 1971, Office of Water Programs, Environmental Protection Agency, Washington DC
Soil Erosion: Causes and Mechanisms: Prevention and Control, 1973, Special Report 135, Highways Research Board, Washington DC
Hydro-seeding, 1989, in Erosion knows no boundaries, Proceedings of 20th Annual Conference of International Erosion Control Association, Steamboat Springs, Colorado

Chapter 13 Streambank stabilization

Streambank stabilization, 1990, chapter 6.4, in The Use of Vegetation in Civil Engineering, editors N. J. Coppin and I. G. Richards, Butterworths and CIRIA

Erosion Control: a Global Perspective, 1991, Proceedings of 22nd Annual Conference of International Erosion Control Association, Steamboat Springs, Colorado

Chapter 13 Reclamation of saline soils

Diagnosis and Improvement of Saline and Alkali Soils, 1954, Agricultural Handbook 60, USDA, Washington DC

Salinity Seminar, Baghdad: Methods of Amelioration of saline and waterlogged soils, 1971, Irrigation and Drainage Paper 7, FAO, Rome

Chapter 13 Reclamation of disturbed land

Geomorphology and Reclamation of Disturbed Land, 1987, T. J. Toy and R. K. Hadley, Academic Press, Orlando, Florida

Surface Mining and the Environment, 1993, M. J. Haigh, International Journal of Surface Mining and Reclamation, 7, 91–104

Chapter 14 National and regional policies

Soil Conservation Policies, Institutions, and Incentives, 1982, edited by H. G. Halcrow, E. O. Heady and M. L. Cotner, Soil Conservation Society of America, Ankeny, Iowa

Conservation Policies for Sustainable Hillslope Farming, 1992, edited by S. Arsyad, I. Amien, T. C. Sheng and W. C. Moldenhauer, Soil and Water Conservation Society, Ankeny, Iowa

Chapter 14 Leglislation

Land degradation in Australia: the search for a legal remedy, 1991, J. W. Looney, Journal of Soil and Water Conservation, 46, 4, 256–259

Chapter 14 Economic analysis

75 Case Studies on Environmental Economic Valuation in Developing Countries, 1993, A. Ekbom, AFTES, World Bank, Washington DC

World Without End: Economic, Environment, and Sustainable Development, 1993, World Bank, Washington DC

Subject Index

Page numbers in *italic* refer to illustrations

Index of names

Page numbers in *italic* refer to illustrations